Learning Materials in Biosciences

Learning Materials in Biosciences textbooks compactly and concisely discuss a specific biological, biomedical, biochemical, bioengineering or cell biologic topic. The textbooks in this series are based on lectures for upper-level undergraduates, master's and graduate students, presented and written by authoritative figures in the field at leading universities around the globe.

The titles are organized to guide the reader to a deeper understanding of the concepts covered.

Each textbook provides readers with fundamental insights into the subject and prepares them to independently pursue further thinking and research on the topic. Colored figures, step-by-step protocols and take-home messages offer an accessible approach to learning and understanding.

In addition to being designed to benefit students, Learning Materials textbooks represent a valuable tool for lecturers and teachers, helping them to prepare their own respective coursework.

Boris Egger

Editor

Neurogenetics

Current Topics in Cellular and Developmental Neurobiology

 Springer

Editor
Boris Egger
Department of Biology
University of Fribourg
Fribourg, Switzerland

ISSN 2509-6125   ISSN 2509-6133   (electronic)
Learning Materials in Biosciences
ISBN 978-3-031-07792-0   ISBN 978-3-031-07793-7   (eBook)
https://doi.org/10.1007/978-3-031-07793-7

This Springer imprint is published by the registered company Springer Nature Switzerland AG
The registered company address is: Gewerbestrasse 11, 6330 Cham, Switzerland

Preface

This collection of chapters aims to accompany a lecture series in neurogenetics. It provides a framework for the discipline that seeks to analyze the relations between the genome, neurons, neural networks, and behavior. The selection of topics in this book aims to provide students with insights into developmental, cellular, and genetic mechanisms in neurobiology. A special emphasis is given to experimental animal model systems such as *Caenorhabditis elegans*, *Drosophila melanogaster*, and vertebrate models. Each chapter starts with an introduction and an overview of the lecture topic. It is followed by the main content, which includes both theoretical and technical experimental parts.

The lecture series addresses students who are doing their master's and doctoral thesis work in the field of neurobiology. It should help and inspire students to establish a theoretical background, think about concepts and design their own research experiments.

Fribourg, Switzerland   Boris Egger

Contents

Introduction to Neurogenetics

1

Boris Egger, Dominique A. Glauser

> **What Will You Learn in This Chapter**
> This first chapter provides an overview of the field of neurogenetics including:
>
> - Specific definitions and examples in order to understand the different meanings that the term *neurogenetics* may have in different fields of science.
> - A brief history of the field of neurogenetic research.
> - A comparative presentation of the main neurogenetic experimental animals and systems, with a specific focus on the strengths of each model.

1.1 Definitions of Neurogenetics

Neurogenetics deals with the relations between the genome and the nervous system. The term *neurogenetics* is used in at least three different ways:

Instrumental or *applied neurogenetics* uses mutants or genetic techniques as tools for analyzing structural or functional aspects of the nervous system. Examples of this are the use of mutants that lead to homeotic transformations of appendages or enhancer trap lines as neuron-type specific markers.

B. Egger (✉) · D. A. Glauser (✉)
Department of Biology, University of Fribourg, Fribourg, Switzerland
e-mail: boris.egger@unifr.ch; dominique.glauser@unifr.ch

© Springer Nature Switzerland AG 2023
B. Egger (ed.), *Neurogenetics*, Learning Materials in Biosciences,
https://doi.org/10.1007/978-3-031-07793-7_1

Analytical neurogenetics attempts to explain the molecular mechanisms of genes that affect the nervous system and elucidate their interactions. It tries to understand how far the nervous system is genetically programmed and whether true "neural genes" exist. Is there evidence for genes whose activity is restricted to the nervous system only? Analytical neurogenetics seeks to elucidate the relations between genes, neurons, neuronal circuits, and behavior.

Medical neurogenetics investigates the genetic background of human neurogenetic disorders (e.g., Microcephaly or Alzheimer's Disease).

1.2   Instrumental Versus Analytical Neurogenetics

The difference between *instrumental* and *analytical* neurogenetics is illustrated in the following examples:

Example for instrumental neurogenetics:

Homeotic mutations in *Drosophila* result in changes in the identity of specific body segments. For example, a *gain-of-function* mutation in the gene *Antennapedia* can lead to an antenna to leg transformation [1]. It raises the question of where sensory neurons from such a transformed appendage project in the brain: to the antennal centers or to the leg centers? Homeotic mutants, therefore, allow us to analyze how the axons of sensory neurons find their correct targets in the brain.

Example for analytical neurogenetics:

Normal visual pathway in mammals: The vertebrate retina is topographically represented in the brain. In mammals, retinal ganglion cells project to the first association area, the lateral geniculate nucleus (LGN) in a topographical manner. The two layers A and A1 receive corresponding projections of the contralateral visual field from both the ipsilateral and the contralateral eye. Secondary fibers extend from the LGN in parallel to area 17 of the visual cortex. Thus, LGN and area 17 both have a binocular representation of the contralateral visual field.

Abnormal visual pathway in albinotic mammals: In albinos (e.g., Siamese cats) A1 receives additional fibers from the contralateral eye (abA1), but less fibers from the ipsilateral eye. Mirror symmetric positions in A1 and abA1 are in the register. In the LGN-cortex projection, the continuity is maintained, i.e., only a small part of area 17 gets binocular information. As a consequence, stereoscopic vision is impaired [2].

In analytical neurogenetics, we ask the question of what is the cause of this neural defect.

It turns out that there is a genetic defect at the *albino* locus. The gene codes for a dopa-oxidase. As a consequence, the synthesis of melanin is blocked. Melanin is used for pigmentation for example in the fur but also in the eye. The loss of pigmentation in the eye stalk leads to impairment of axon growth. The direction of growing axons into the optic chiasm is modified. Therefore, in the LGN and in the visual cortex the visual input is represented in an abnormal manner. The animals display impaired stereoscopic vision.

This example shows that the activity of a non-neural gene can induce a cascade of events that finally leads to specific structural and functional defects in the nervous system. It illustrates that many genes fulfill different tasks during different periods of development, within and outside the nervous system. Nevertheless, many neural-specific genes have been described as we will learn in subsequent chapters. It is estimated that in mammals around 80% of genes encoded by the genome are expressed in the central nervous system [3]. Many of these genes have a broad spectrum of expression inside and outside the nervous system and might be active in specific developmental periods. Only a few genes might be truly neural-specific genes and are only required in neuronal and glial subtypes at later stages of development or during adulthood.

1.3 History of Neurogenetics

It is difficult to define the origins of neurogenetics. However, we can clearly find concepts and designs of experimental studies expanding in the 1960s–1970s, which tried to elucidate the connection between genes and behavioral phenotypes. Most pioneering minds in the field of neurogenetics belonged to Sydney Brenner at the Medical Research Council in Cambridge, UK [4] and to Seymour Benzer at the California Institute of Technology [5]. The quotes below present some points of view illustrating the need and promises of the implementation of neurogenetics approaches and how they could help obtain a comprehensive understanding of the nervous system structure and function.

> The problem of tracing the emergence of multidimensional behavior from the genes is a challenge that may not become obsolete so soon. (Seymour Benzer).

> A thorough analysis of the effects of such mutations might throw light on the logical structure of the programme. As to how the effects of such genes are mediated is an entirely separate question at the moment. (Sydney Brenner).

> Behaviour is the result of a complex ill-understood set of computations performed by nervous systems and it seems essential to decompose the question into two: one concerned with the question of the genetic specification of nervous systems and the other with the way nervous systems work to produce behaviour. (Sydney Brenner).

> So genocentric has modern biology become that we have forgotten that the real units of function and structure in an organism are cells and not genes. (Sydney Brenner).

A chronology of some selected important landmarks and periods in the history of the neurogenetics field:

- 1950s: behavioral mutants of *Drosophila melanogaster* (Seymour Benzer, Caltech)
- 1960s: characterization of locomotor mutants of mice (Sidman and others)
- 1960–70: establishment of the *C. elegans* model for neurogenetic studies (Sydney Brenner, MRC Cambridge, Nobel Prize 2002)

- 1970–80: expansion of the discovery of neurological and behavioral mutants in *Drosophila*, *C. elegans* and mouse
- 1977: first report of a neurological illness mapping to a specific human chromosome (spinocerebellar ataxia type 1, Jackson and colleagues)
- 1980–90: first behavioral mutants in Zebrafish (Streisinger and others)
- 1980–present: molecular genetic analysis of many neural genes: discovery of homologous neural genes, conserved neural pathways
- 1986: report of the full *C. elegans* connectome, (White, Brenner and colleagues)
- 1994: first report of GFP expression in the nervous system of *C. elegans* (Chalfie and colleagues)
- 1990–present: development of genetically encoded calcium indicators for the recording of neural activity (Tsien, Looger, and others)
- 1998–present Whole-genome sequences are available for many model organisms, which allow genomic screens for developmental or behavioral mutants.
- 2000s: development of optogenetic tools for the manipulation of neural activity (Tsien, Deisseroth, Raymond, Miesenbock, Bamberg, Henneman, and others)
- 2007: Brainbow (Lichtman, Sanes and collaborators)
- 2012: The Allen Brain Atlas (Jones and colleagues)
- 2013 whole brain calcium imaging in larval Zebrafish with recordings at cellular resolution (Ahrens, Keller, and colleagues)
- 2013–present: retinal and cerebral organoid models to study human brain diseases (Sasai, Lancaster, Knoblich and colleagues)
- 2017–present: connectomes of major brain learning centers in the *Drosophila* brain (Rubin, Waddell, Plaza, Cardona, Bock and colleagues)

1.4   Model Systems in Neurogenetics

Only a few neurogenetic model systems have emerged to this date (Fig. 1.1). The small number is mainly due to the stringent requirements in breeding and experimenting with model organisms. The ideal model organism is easy to breed and has a short life cycle and hence generation time. Among other advantages, this enables the researcher to experiment with a large number of progenies. The ideal model organism also possesses a small genome size, which simplifies genetic experiments. It allows the use of powerful genetic and molecular techniques. Ideally, it reveals a small transparent body and has a limited number of neurons to uniquely identify cellular lineages and neuronal circuits. Depending on the biological questions asked it should also have an interesting behavioral repertoire, which is relevant for nervous system function in higher species. Therefore, the finding might also be relevant for medical neurogenetics. With *Caenorhabditis elegans* and *Drosophila melanogaster* two classical invertebrate model organisms are listed, which contributed tremendously to our current understanding of how a bilaterian nervous system develops and functions. They are followed by vertebrate models with more complex nervous

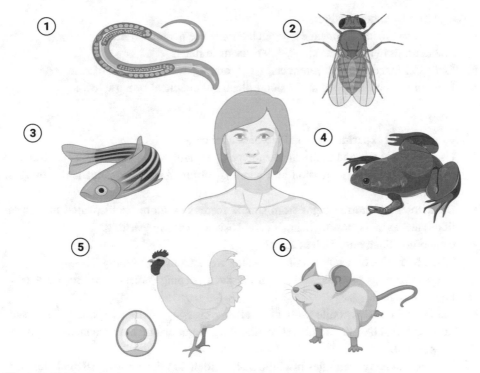

Fig. 1.1 Six major animal model organisms used in neurogenetic research. The genetic and neurobiological characteristics of the following model organisms are described in more detail: (1) *Caenorhabditis elegans* (nematode worm), (2) *Drosophila melanogaster* (fruit fly), (3) *Danio rerio* (zebrafish), (4) *Xenopus laevis* and *Xenopus tropicalis* (African claw frog), (5) *Gallus gallus* (chicken), (6) *Mus musculus* (mouse), and *Homo sapiens* (human). Created with BioRender.com

systems. Each of these model systems is particularly suited to address specific questions in the field of neurogenetic research. Finally, in the past decade, great advances were made in establishing human organoids in order to study the normal development of the human brain or to model neuro-developmental diseases. The following paragraphs summarize key features of these different models in a synthetic format and are based on multiple sources [6–10].

1.4.1 *Caenorhabditis elegans*

Genes and genomes:
- Six chromosomes (five autosomes (A) and one sex chromosome (X)).
- Sex determination: 5AA and XX in hermaphrodite; 5AA and X0 in males.
- ~100 MB genome, entirely cloned and sequenced
- ~20,200 coding genes

Neurons and nervous system:
- Very stereotyped cell lineage tree, including in the nervous system.
- 302 neurons in hermaphrodites, 359 neurons in males
- Only 150 functionally different cell types, among them 118 types of neurons.
- Full connectome mapped at the single electrical/chemical synapse level.

Key features for experimenters and research applications:
- Used to study a broad range of phenotypes, including behavior, circuit function and development, gene expression and neuronal differentiation, cellular and molecular physiology.
- Behavioral repertoire ranges from simple feeding to more sophisticated manifestations such as taxis, mating behavior and basic associative learning.
- Generation time: only 52 h at 25 °C.
- Small body size of 1 mm length allows for breeding in high numbers.
- Deep-freezing is possible and simplifies stock keeping (preservation for decades in liquid nitrogen).
- Mating of hermaphrodite with male allows crossing of mutant strains. The self-fertilization of hermaphrodites simplifies the generation and maintenance of homozygous strains.
- Transparent body simplifies brightfield and widefield microscopy, laser cell ablation, and optogenetics.
- A large collection of neurological mutants is available.
- Comprehensive quantification of behavior thanks to computer-assisted analytic platforms.
- Web-resources: wormbase.org

1.4.2   *Drosophila melanogaster*

Genes and genomes:
- Three autosomal (A) and sex chromosomes (X/Y).
- ~143 MB genome, entirely cloned and sequenced
- ~13,900 coding genes (ensembl.org)

Neurons and nervous system:
- Neuronal and glial lineages are mapped and described in great detail.
- Nervous system consists of over 100,000–200,000 neurons in the adult fly, 10,000–15,000 in larvae.

Key features for experimenters and research applications:
- Meiotic recombination is suppressed in males.
- Used to study a broad range of phenotypes, including behavior, circuit function and development, gene expression and neuronal differentiation, cellular and molecular physiology.
- Displays a large behavioral repertoire including different locomotion modes and social interactions.
- Generation time is about 10 days at 25 °C.
- Polytene chromosomes reveal a characteristic pattern of 102 major bands and allow easy identification of chromosomal deletions and rearrangements.
- Classical and molecular genetic applications are available.
- Genome-wide and tissue-specific gene knockdown is possible through inducible RNAi lines.
- Cell autonomous gene function can be assessed through sophisticated mutant clonal systems such as the MARCM (mosaic analysis with a repressible cell marker) system.
- Many non-genetic manipulations such as neurophysiology, pharmacology, *in vitro* culture and microscopy are available.
- Web-resources: flybase.org

1.4.3   Vertebrate Animal Models

Zebrafish (*Danio rerio*)

Genes and genomes:
- 25 chromosomes
- ~1.7 GB genome, entirely cloned and sequenced
- ~25,600 coding genes

Neurons and nervous system:
- Adult zebrafish has about 10^7 neurons, about 80,000 in a one-week-old larva.
- Nervous system has large identifiable neurons.

Key features for experimenters and research applications:
- Classically used for neuro-developmental studies, including for modeling of neuro-logical disorders; more recent emergence of functional studies including whole-brain activity imaging.
- Generation time is about 3–5 months.
- Brood size: 200–300.
- Efficient techniques for the transient manipulation of gene activities in embryo.

- Transparent skin and therefore excellent model organism for live cell imaging.
- Large behavioral repertoire, easy to track and quantify.
- Web-resources: zfin.org

African clawed frogs (*Xenopus laevis* and *X. tropicalis*).

Genes and genomes:
- 18 chromosomes in *X. laevis;* 10 chromosomes in *X. tropicalis*
- 3.1 Gb in *X. laevis*; 1.7 Gb in *X. tropicalis*, both entirely sequenced
- *X. tropicalis* contains an estimated 20,000 coding genes (ensembl.org).

Neurons and nervous system:
- ~10^8 neurons in adults; a few thousand in hatchling tadpoles.

Key features for experimenters and research applications:
- Generation time of 1–2 years for *X. laevis* or 4 months for *X. tropicalis*.
- Large brood size 500–3000.
- Xenopus oocytes are very large cells and easy to culture and to inject.
- *Xenopus laevis* was used for the first nuclear transfer or cloning experiments.
- Most suitable to study the early development and patterning of the nervous system.
- Growing behavioral repertoire within a few days in tadpoles.
- Web-resources: xenbase.org

Chicken (*Gallus gallus*)

Genes and genomes:
- 39 chromosomes
- ~1.0 Gb genome, entirely cloned and sequenced
- Contains an estimated 16,900 coding genes (ensembl.org).

Neurons and nervous system:
- ~10^8 neurons in adults (estimation based on Red Junglefowl, a relative of the domesticated chicken)
- Conservation of most brain regions in comparison to human (basal ganglia, amygdaloid nuclei, hippocampus, hypothalamus, isocortex).

Key features for experimenters and research applications:
- Amniote and therefore easy to manipulate (i.e., electroporation *in ovo*, *ex ovo* culture).
- Good model for studies on neuronal cell type specification and on axon guidance.

Mouse (*Mus musculus*)

Genes and genomes:
- 19 paired autosomal chromosomes and X/Y sex chromosomes
- ~3.5 Gb genome, entirely cloned and sequence
- Contains an estimated 22,500 coding genes (ensembl.org).

Neurons and nervous system:
- The mouse brain contains about 70 million neurons.
- Brain region organization is similar to human.

Key features for experimenters and research applications:
- Major mammalian model system to study developmental genetics and neurogenetic disorders.
- Advanced molecular and genetic tools are available (i.e., homologous recombination, sophisticated clonal tracing systems).
- Many mice-derived neuronal cell culture systems are available.

1.4.4   Human (*Homo sapiens*)

Genes and genomes:
- 22 paired autosomal chromosomes and X/Y sex chromosomes
- ~4.5 Gb genome, entirely cloned and sequenced
- Contains an estimated 20,400 coding genes (ensembl.org).

Neurons and nervous system:
- The human brain contains approximately 86 billion neurons.

Key features for experimenters and research applications:
- Human-derived cell culture systems are available.
- Major recent advances stem cell and organoid technologies allow to model neurodevelopmental disorders (i.e., microcephaly, neural tumor biology).

Take-Home Message

- Seymour Benzer and Sydney Brenner were two of the pioneering minds in the field of experimental neurogenetics.
- Neurogenetic research can be classified into instrumental, analytical, and medical.
- A good research model system needs to fulfill several criteria for optimized breeding, tractability of the genome, and accessibility of the nervous system.

Question

Explain the differences between instrumental and analytical neurogenetics?

Acknowledgments We would like to thank Professor Emeritus Dr. Reinhard Stocker, for contributions to this chapter.

References

1. Wagner-Bernholz JT, Wilson C, Gibson G, Schuh R, Gehring WJ. Identification of target genes of the homeotic gene Antennapedia by enhancer detection. Genes Dev. 1991;5(12B):2467–80.
2. Shatz CJ, LeVay S. Siamese cat: altered connections of visual cortex. Science. 1979;204(4390):328–30.
3. Hawrylycz MJ, Lein ES, Guillozet-Bongaarts AL, Shen EH, Ng L, Miller JA, et al. An anatomically comprehensive atlas of the adult human brain transcriptome. Nature. 2012;489(7416):391–9.
4. Brenner S. The genetics of *Caenorhabditis elegans*. Genetics. 1974;77(1):71–94.
5. Benzer S. From the gene to behavior. JAMA. 1971;218(7):1015–22.
6. White BH. What genetic model organisms offer the study of behavior and neural circuits. J Neurogenet. 2016;30(2):54–61.
7. Roberts A, Li WC, Soffe SR. How neurons generate behavior in a hatchling amphibian tadpole: an outline. Front Behav Neurosci. 2010;4:16.
8. Corsi AK, Wightman B, Chalfie M. A transparent window into biology: a primer on *Caenorhabditis elegans*. Genetics. 2015;200(2):387–407.
9. Tsapara G, Andermatt I, Stoeckli ET. Gene silencing in chicken brain development. Methods Mol Biol. 2020;2047:439–56.
10. Kim J, Koo BK, Knoblich JA. Human organoids: model systems for human biology and medicine. Nat Rev Mol Cell Biol. 2020;21(10):571–84.

Further Reading

White BH. What genetic model organisms offer the study of behavior and neural circuits. J Neurogenet. 2016;30(2):54–61.

Neurogenetic Analysis in *Caenorhabditis elegans*

2

Saurabh Thapliyal, Dominique A. Glauser

What You Will Learn in This Chapter
The goal of this chapter is to introduce the *Caenorhabditis elegans* model and how it can be used to address fundamental questions in neurobiology using neurogenetics. In the first part, we will review the specific features and advantages of the *C. elegans* model and highlight why it is particularly suited for systems biology approaches to bridge our understanding at the molecular/genetic, cellular, circuit, and behavioral levels. In the second part, we will present how *C. elegans* can be used for the functional analysis of genes and neural circuits. Covered topics will range from very classical approaches (such as forward genetics and reporter genes) to more modern methods (such as optogenetics and single-cell RNAseq). The goal of this second section is not to provide hands-on protocols, nor to be exhaustive, but to present an overview of some common strategies, their principle, their usefulness, and some of their limitations. Finally, in the third part, we elaborate on the current challenges and future developments in the use of the *C. elegans* model for neurogenetics. This chapter should help readers grasp why neurobiological research with *C. elegans* has been and is still so successful and serve as a kick-off for newcomers in *C. elegans* neurogenetics by easing their access to research literature.

S. Thapliyal · D. A. Glauser (✉)
Department of Biology, University of Fribourg, Fribourg, Switzerland
e-mail: saurabh.thapliyal@unifr.ch; dominique.glauser@unifr.ch

© Springer Nature Switzerland AG 2023
B. Egger (ed.), *Neurogenetics*, Learning Materials in Biosciences,
https://doi.org/10.1007/978-3-031-07793-7_2

2.1   *C. elegans* Model for Neurobiology

C. elegans is a very important model for biology. Researchers isolated and started to study it more than a century ago [1]. In the 1960s, molecular geneticist Sydney Brenner decided to turn to this model, as it seemed ideal for deciphering the link between gene and animal behavior. Since this turning point, continuously growing research with *C. elegans* has brought a large volume of knowledge about the molecular, cellular, and circuit bases of the nervous system development and functions [2]. In this first part, we will summarize the key features of the model.

2.1.1   A Powerful Genetic Model

C. elegans was the first animal to have its genome entirely sequenced [3]. The worm genome is relatively small, about 100,000 kb. It contains about 20,000 genes, spreading over 6 chromosomes: 5 autosomes (chromosome I, II, III, IV, and V; existing as chromatid pairs in diploid cells) and 1 sexual chromosome (chromosome X, existing as a chromatid pair in hermaphrodites or as a single chromatid in males). The hermaphrodite's ability to self-fertilize makes the generation and maintenance of homozygous strains very convenient. Yet, genetic crossings involving hermaphrodites and males are still possible. Other advantageous aspects of this model include a fast generation time (it takes, for example, only 52 hours at 25 °C for one generation) and the possibility to deep-freeze strains in liquid nitrogen, which simplifies the keeping of genetic stocks for later use. The worms can grow in well-controlled culture conditions, usually fed with *E. coli* bacteria. The animals being extremely small (~1 mm in length for an adult), large numbers of strains and animals can be grown in relatively small volumes and at reasonable costs. In addition, transgenic lines can be obtained relatively quickly in comparison to other models. Typically, it can take as little as 10 days, depending on the type of transgenic animals (see more details under Sect. 2.2.2) [4], whereas transgenesis in mice can take several months.

2.1.2   Evolutionary Conservation of Neuronal Genes

Whereas estimates vary depending on the comparison method used, it was estimated that 60 to 80% of human genes have an ortholog in *C. elegans* [5] and 40% of genes associated with human diseases have a *C. elegans* ortholog gene [6]. Homologous genes are found for a majority of neuronal genes [see the comprehensive review in 7]. Among conserved genes, some encode ion channels and gap junctions, components of neurotransmission pathways, neuropeptides, intracellular signaling components (e.g., G-protein coupled receptors, cGMP pathway), neuronal recognition, and adhesion molecules. Voltage-gated sodium channels and HCN channel homologs have not been reported to date in *C.*

elegans, two notable exceptions in this largely conserved landscape. Consistent with the absence of voltage-gated channels, worms do not produce sodium-based action potentials, but calcium-based action potentials [8]. Apart from this electrophysiological peculiarity, the development, ultrastructure, and functions of worm neurons present similarities with those of other animals. Worms notably use a similar set of neurotransmitters for synaptic transmission (e.g., acetylcholine, GABA, glutamate), biogenic amines (dopamine, serotonin, and octopamine) for synaptic and para/endocrine signaling, as well as a rich set of neuropeptides for neuromodulation [9]. The latter form of cell-cell communication appears to be particularly well-developed in *C. elegans*, as is often the case in invertebrates.

2.1.3   The Architecture of the *C. elegans* Nervous System

The neuro-anatomy of *C. elegans* is particularly well-described [10]. The animal develops with a stereotyped cell lineage and neuronal cell bodies form an architecture that is very similar from animal to animal. Of the 959 somatic cells in hermaphrodites, 302 are neurons. Males contain 79 additional neurons, most of which are located in the tail and take part in the relatively sophisticated worm reproductive behavior. Figure 2.1 presents a schematic view of a hermaphrodite nervous system that can be divided into two distinct units: the pharyngeal nervous system (encompassing 20 neurons, with functions mostly dedicated to the regulation of food intake) and the somatic nervous system (282 neurons). The two systems are connected only at the level of a single interneuron pair. The main *C. elegans* neuropil (a region of densely interconnected neural fibers) is called the nerve ring and forms a loop around the pharynx. The majority of neuron cell bodies are clustered in a few ganglia adjacent to the nerve ring in the head of the animal, together forming a brain-like structure. A large number of ciliated sensory endings stem from this region to reach the tip of the nose in a structure called the amphid. A similar, yet smaller, structure is found in the animal tail: the phasmid. The ventral and dorsal nerve chords, as well as some lateral tracks (including, e.g., the neurite of mechanosensory neurons), run along the animal body.

Neurons can be categorized based on known functional properties and their location in the neural network as sensory neurons, interneurons, motor neurons, and modulatory neurons. These designations are somewhat arbitrary and many neurons may be assigned

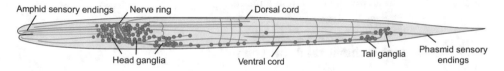

Fig. 2.1 Schematic of the *C. elegans* nervous system. Circles represent neuron cell body positions. The pharyngeal nervous system is in orange and the somatic in blue. The nervous systems follow a bilateral symmetry (only one side depicted)

to more than one category, in which case they are sometimes referred to as "polymodal neurons." The latter term should not be confounded with "polymodal sensory neurons," which refers to sensory neurons able to detect stimuli of different sensory modalities (e.g., chemical and mechanical stimuli). Of about 150 functionally different cell types that have been defined in *C. elegans*, 118 are neuron types. A unified nomenclature has been established such that each neuron can be identified with a 2–3-character name, followed by L (left), R (right), D (dorsal), or V (ventral). The Wormatlas [10] is a useful resource as a starting point for getting an overview of the diversity of *C. elegans* neurons and learning about individual ones. Most neurons exist in two bilateral copies (left and right) and fulfill redundant functions. But there are well-documented exceptions to this rule, where the left and right neuron functions diverge, such as the ASER and ASEL salt-sensing neurons, which detect different ions [11].

2.1.4   The *C. elegans* Synaptic Connectome

The synaptic connectome (assembly of all synaptic connections) of *C. elegans* was fully mapped thanks to the laborious work of *C. elegans* model pioneers in the 1960s–1980s [12]. This connectivity diagram has been an invaluable resource for guiding research on circuit function. However, because the first connectome data were obtained with only two animals, with only partial overlap between them, it was not initially possible to assess with certainty how the nervous system architecture varied from one animal to another. Recently, the *C. elegans* connectome was revisited, expanding the analysis to males [13], as well as to several animals during the post-embryonic development in the nerve ring region [14], providing insights on the connectome differences between sexes, developmental stages, and individuals. Figure 2.2 presents the connectivity diagram of the adult hermaphrodite nervous system and its connection to major muscle groups in a top-down manner following the flow of information. This scheme summarizes around 6400 chemical synapses, 900 gap junctions, and 1600 neuromuscular junctions. The results of recent developmental analyses have provided much important missing information regarding nervous system development and animal-to-animal variations [14]. The overall shape of the nervous system is maintained across development and animals, and some major connection patterns (corresponding to ~70% of all synapses) seem to be maintained. However, an important part of the connectivity is reshaped during development and may undergo different trajectories in different animals.

2.1.5   *C. elegans* Behavioral Repertoire

Beyond simple behaviors, such as locomotion, feeding, and defecation, worms also display more sophisticated behaviors [15]. For example, worms adjust their behavior according to features of their environment, responding to changes in temperature, oxygen, carbon

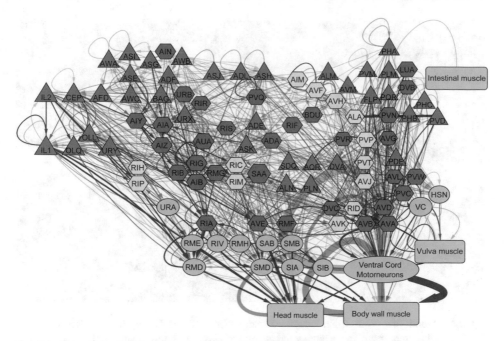

Fig. 2.2 Schematic diagram of the somatic nervous system connectome of an adult hermaphrodite. Synaptic connections (black arrows), gap junctions (red lines), sensory neurons (pink triangles), interneurons (red hexagons), modulatory neurons (yellow hexagons), motor neurons (blue ovals), muscle (green rectangles). Some motor neurons, muscle groups, as well as bilaterally symmetrical neurons are aggregated. The display was generated with Cytoscape (https://cytoscape.org/) from the data by Cook and collaborators [13]

dioxide, salt, and osmolarity. Animals can navigate toward food sources and detect attractive and repulsive odorants to produce chemotaxis. They are able to respond to other individuals via pheromone signaling and each sex also produces sex-specific behaviors, such as egg-laying in hermaphrodites and copulation behavior in males [16]. Most of these behaviors are subject to plasticity and modified by learning and memory. Food availability is an extensively studied factor involved in the modulation of these responses and has been shown in many instances to involve serotonin and/or neuropeptide signaling. *C. elegans* has also been used as an exquisitely simple model of decision-making, to clarify the molecular, cellular, and circuit basis of this process [17–20].

Over the years, researchers have been remarkably creative in developing assays to analyze *C. elegans* behavior. These include a variety of procedures such as the manual scoring of single animal behavior under a stereomicroscope (e.g., to record responses to acute mechanical, chemical, or thermal stimuli), larger-scale population-based assays in which the worm distribution is scored in anisotropic environments (e.g., with thermal- or chemical gradients, thermal or chemical barriers), or methods to assess the impact of pharmacological agents *in vivo* (e.g., animal paralysis in response to neurotransmission-

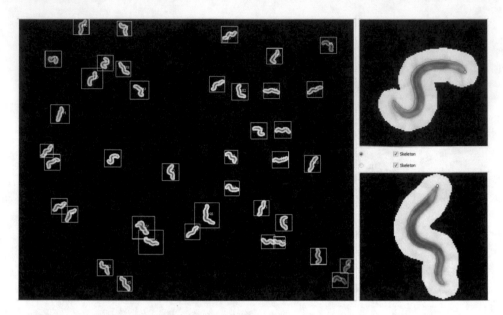

Fig. 2.3 Quantifying postural and locomotion parameters with the Tierpsy tracker. Screenshot of a Tierpsy tracker viewer window showing the parallel tracking of multiple worms (large image, left) and single worm details (insets, right). Animal nose tip (white dot), skeleton (green line), and contour (red line) are scored for every frame of a movie recording adult worm crawling behavior

affecting drugs) [21, 22]. With the development of machine-vision approaches, computer-assisted analyses have emerged to produce high-content behavioral data with automated tracking systems [23]. The most recent systems, such as the Tierpsy tracker [24], can track multiple worms at high resolution to provide a quantitative readout of their locomotion and postural phenotypes (Fig. 2.3). Overall, the behavioral repertoire of *C. elegans* is both relatively rich and accessible for a comprehensive quantitative description.

2.1.6 Major Strengths of the Model

The two main strengths of the *C. elegans* model (Fig. 2.4) are that (i) it allows researchers to gather systematic information at the molecular, cellular, circuit, and behavioral levels, but also (ii) it bridges these different levels. These features are the key to integrative studies that can shed light on processes that often remain closed black boxes in other models.

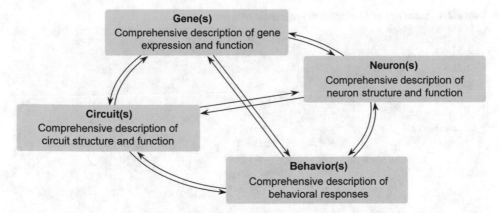

Fig. 2.4 An illustration of the two main strengths of the *C. elegans* model. The ability to perform systematic analyses at multiple levels (gray rectangles), and the capacity to bridge the biological functions across these levels (arrows)

2.2 Experimental Approaches for the Neurogenetic Analysis of Genes and Neural Circuits in *C. elegans*

C. elegans has flourished as a powerful model system to understand how genes control behavior, which is not a trivial task. It requires to elucidate many steps in the pathway from genes to behaviors:

- Identify the genes required for a given behavior or neuronal phenotype (2.2.1)
- Determine the expression pattern of neuronal genes (2.2.2 and 2.2.3)
- Determine in which neuron(s) a specific gene (and its product) play a functional role (2.2.4)
- Understand how these neurons act in a given circuit to control behavior (2.2.5)

This section covers some of the most common experimental strategies used to address these questions and summarized in Table 2.1.

2.2.1 Genetic Screens

C. elegans is a particularly powerful model for identifying genes required for specific phenotypes. Two approaches have mainly been used: (i) mutagenesis screens (classical forward genetics) and (ii) RNA*i* screens. In the latter case, the reverse genetics RNA*i* technique can be used on a very large scale, actually covering the majority of the worm genes. Hence, a large-scale RNA*i* screen can be considered as a form of forward genetics, where researchers start with a phenotype and subsequently determine the causative

Table 2.1 Summary of *C. elegans* neurogenetic tool-kit

Gene analysis	Neuron and circuit analysis	Behavior analysis
Gene identification: • Mutagenesis or RNAi screens Gene expression pattern: • Reporter analysis • Neuron-specific transcription profiling (PAT-seq, TRAP, FACS-seq, RAPID) Gene locus of action determination: • Neuron-specific rescue • Neuron-specific knockdown (RNAi) or knockout (CRISPR-Cas9)	Monitoring neuronal activity: • Calcium indicators (GCaMP, Cameleon) Neuronal ablation: • Laser-ablation, expression of caspase or MiniSOG Modulating neuronal activity: • Enhancing neurotransmission (PKC-1 gain-of-function). • Inhibiting neurotransmission (Tetanus toxin) • Upregulating neuronal activity (ChR2 or Chrimson) • Downregulating neuronal activity (chemogenetics with Histamine-gated chloride channels; optogenetics with Halorhodopsin, Archeorhodopsin or Anion Channelrhodopsins)	Behavior quantification: • Large catalog of standardized behavioral assays • Tierpsy tracker for high-content computer-assisted behavior quantification

genetic alteration [25]. Importantly, the outcome of a genetic screen is usually only a starting point, since deciphering how the different identified gene products control specific neuronal/behavioral phenotypes requires additional experiments.

2.2.1.1 Mutagenesis Screen
Principle
Mutagenesis screens have been and are still extensively used to identify genes controlling neuron development and function [26, 27]. The principle is to mutagenize a population of worms, generally using a chemical reagent (e.g., EMS or ENU), and screen the resulting mutants for specific phenotypes [28]. These could be, for example, phenotypic defects in the execution of a given behavior, in the morphology of neurons, or in neuronal gene expression.

Mutation Mapping and Validation
Once mutant lines are recovered, the mutations need to be mapped and the causal sequence changes identified. While the mutation mapping used to be an extremely tedious process, the recent implementation of Whole-Genome Sequencing (WGS)-based approaches considerably accelerated this limiting step [29, 30]. Because chemical mutagenesis typically causes several dozen gene-coding mutations in each mutant line, obtaining the genomic sequence of a mutant is usually not sufficient for the identification of a causative mutation. In order to distinguish between a causal mutation and non-causal side-

mutations, additional upstream and/or downstream work is required, which may include gross mutation mapping, extensive backcrossing, and/or the preparation of pooled variant batches prior to WGS [31]. Once a mapping interval has been narrowed to include only one (or a few) candidate mutation(s), it is necessary to confirm that the candidate mutation(s) indeed cause(s) the phenotype, using one or more of the following approaches:

i. Complementation test (for recessive alleles, when another loss-of-function allele is available)
ii. Genetic rescue experiments (e.g., by transforming a genomic fragment)
iii. Phenocopy test, to verify that the same phenotype is present in additional available mutants, upon gene knockdown with RNA*i*, or (ideally) in a CRISPR/Cas9 genome-edited mutant recreating the same mutation

Mutagenesis Screen Example: Mechanosensation

The screens performed to uncover the molecular substrates controlling mechanosensory behavior represent a very interesting example of a *C. elegans* forward genetics success story. Worms detect gentle touch stimuli thanks to six touch receptor neurons: AVM, ALML/ALMR, PVM, and PLML/PLMR (Fig. 2.5a) [32]. These neurons send processes along the animal body, where gentle touch stimuli are detected in a process called mechanotransduction (i.e., the transformation of mechanical force into electrical signals). The neurites of these neurons have two distinctive ultrastructural features: an intracellular bundle of large-diameter microtubules and a prominent electron-dense extracellular "mantle" that anchors the axon to the cuticle. Both elements may serve as stimulus-transducing structures. By screening for mutants insensitive to gentle touch using a simple behavioral screen (touching anterior or posterior ends of the worm with an eyebrow hair), more than 450 gentle touch insensitive mutants were isolated, defining 18 relevant genes [33]. The screening procedure also included a step to filter out mutants unable to respond to a harsh touch (pocking with a platinum wire). Because multiple alleles were isolated for all genes (except for the weak phenotype mutant *mec-17*), this screen is probably saturated for genes whose loss-of-function selectively confers a gentle touch insensitivity. Most mutants are called *mec* for *mec*hanosensation defective, as they were first identified based on this phenotype. As illustrated in Fig. 2.5b, *mec* mutants affect many processes essential for the development and function of the touch receptor neurons, which has been revealed in numerous follow-up studies. Affected genes include those essential for the cell lineage (*unc-86* and *lin-32*) and differentiation of touch receptor neurons (*mec-3*, coding for a terminal differentiation transcription factor), those essential for the formation of extra- and intracellular structures in the touch receptor neurons (*mec-1*, -5, -7, -9, -12, -17), as well as those coding for components of the mechanotransduction channel complex (*mec-2*, -4, -6, -10). Overall, the mechanosensation forward genetics example illustrates the large variety of biological pathways and molecular actors that can be enlightened with behavioral defect-based screens and subsequent analyses.

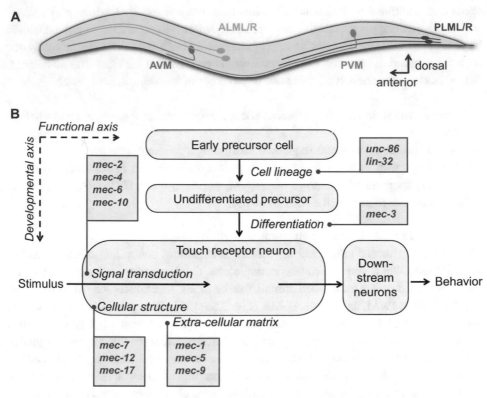

Fig. 2.5 Neurons and genes required for gentle touch response. Schematic of the worm showing the localization of touch receptor neurons (**a**). Diagram illustrating the many processes (italic) controlled by the *mec* genes (red), whose mutations affect gentle touch response (**b**)

2.2.1.2 Large-Scale RNAi Screen

RNA interference (RNA*i*) was first discovered in *C. elegans* [34]. RNA*i* can be induced by microinjection of double-stranded RNA (dsRNA) or dsRNA-encoding DNA, by soaking the animal in dsRNA solutions, or by feeding with specific *E. coli* lines expressing dsRNAs [35]. The latter approach could be up-scaled to perform genome-wide screens. To this end, libraries of *E. coli* strains covering a majority of the *C. elegans* genes have been created [36]. Each *E. coli* strain expresses a specific dsRNA and will knock down the corresponding *C. elegans* gene when worms are fed with that specific bacterial strain. Therefore, RNA*i*, which is most commonly used for reverse genetics (where a known gene is first knocked down and the resulting phenotype is subsequently characterized), can thus be used in a forward genetics setting [25].

Screening an almost entire genome requires growing ~18,000 different bacterial strains, feeding worms on each bacteria culture, and scoring the given phenotype for each of them. The screening steps involve more work than a typical mutagenesis screen [37]. However, unlike mutagenesis screens, where the identification of the causal mutation is

Table 2.2   Distinctive features of mutagenesis and RNAi-based screens

Mutagenesis screen	RNAi screen
Both loss- and gain-of-function alleles can be recovered	Only gene expression knockdown (no gain-of-function, no full knockout)
Point mutations can bring insights into structure-function relationships	Multiple isoforms and genes with shared sequences may be knocked down (redundancy bypass)
Every gene is potentially targeted (unbiased approach)	Unequal susceptibility to RNAi of different genes (protein-half-life) or tissues (RNAi-resistant tissues)
Mutations affect single genes, potentially single isoforms	
Screening is typically less tedious than for RNAi screens, but mutation mapping is time-consuming and costly	RNAi can be triggered after initial development (bypassing early requirement)
	Screening at a genomic scale is tedious and costly, but the identity of target genes is known immediately

a major bottleneck, the identity of the knocked-down gene is immediately known in RNA*i* screens, constituting a major advantage. The fact that RNA*i* generally produces a partial knockdown and that RNA*i* can be induced in a delayed manner (e.g., by transferring animals at a post-embryonic stage on RNA*i* *E. coli*) may represent another advantage in the study of genes whose permanent knock-out is lethal. However, RNA*i* screens lack the ability to produce full knock-out or gain-of-function via point mutations and to produce mutant lines that can be used for further studies. Furthermore, some genes are refractory to RNA*i*. An additional complication for the application of RNA*i* to *C. elegans* neurobiology is that the feeding RNA*i* approach has a reduced efficacy in the nervous system of the classical wild-type (N2) animals. Specific genetic backgrounds must be used in this case, including specific RNA*i*-enhancing mutants (e.g., *eri-1; lin15b*) and/or transgenic animals expressing the SID-1 RNA transporter in neurons, in order to help dsRNA molecules reach the nervous system [38].

Table 2.2 presents a summary of the main differences between mutagenesis and RNA*i* screens.

2.2.2   Transgenic Approaches in *C. elegans*

Many modern neurogenetic analyses are performed with transgenic animals, including the majority of the methods presented in the rest of this chapter. It is thus important to understand how transgenesis is performed in *C. elegans* and how transgene expression in target cells can be achieved.

2.2.2.1 Transgenesis in *C. elegans*

To obtain transgenic worms, ectopic DNA, usually plasmids, fosmids, or PCR products, must be transferred into the animal germline [39, 40]. Two methods are used: micro-injection of DNA solution directly into the gonad or bombardment (biolistic approach) with gold particles coated with DNA. Following DNA transformation, and depending on the nature of the injected material and the subsequent procedure, different types of transgenic animals can be created, as elaborated below and summarized in Table 2.3. Some resulting transgenic animal types are quite different from what exists in other model organisms and their particularities must be understood in order to properly interpret the results obtained with transgenic worms.

Extrachromosomal Transgene Array

After DNA transfer into the germline of a treated P_0 population and after egg laying, some of the progeny (F_1 generation) will have retained the DNA. Typically, the ectopic DNA is not integrated into the animal chromosomes and retention is unstable, such that most F_1 animals will not be able to replicate/transmit the transgene to subsequent generations. However, a portion of those F_1 animals (~10–30%, but this can vary considerably) will produce transgenic progeny when allowed to self-fertilize and the transgene will be inherited in subsequent generations. In this case, the DNA forms what is called an *extrachromosomal transgene array* (or *extrachromosomal array*): an assembly of several DNA copies (>100) that is replicated and transmitted during cell divisions, but that is not part of any chromosome [41, 42]. The transmission is not perfect from cell to cell. Animals are mosaics with some cells carrying the transgene and others not. This is true also for the germ cells. Thus, the transgene transmission to the next generation is also unperfect (unlike chromosomal DNA). The resulting transgenic lines can be perpetuated and frozen and are referred to as *stable extrachromosomal array lines* or *stable lines*. The maintenance of stable lines requires a selection co-marker. Commonly used markers include the *Rol(+)* transgene inducing a *roller* phenotype (which may interfere with some behavior analyses), fluorescent proteins expressed in a variety of tissues (requiring maintenance under a fluorescent scope), as well as the rescue of growth-impairing mutations (which will require the use of a specific mutant background, e.g., the temperature-sensitive *pha-1* allele) [41, 43, 44].

Integrated Arrays

Extrachromosomal arrays can be integrated into chromosomes, resulting in Mendelian inheritance of the transgene (with 100% transmission rate in homozygotes) and solving the problems associated with mosaicism. Integrated arrays retain multiple copies of the transgene, which is advantageous when a strong transgene expression is needed. Spontaneous integration of extrachromosomal arrays is extremely rare after gonad micro-injection of DNA, but much more common after bombardment [39, 45, 46]. Integration of stable extrachromosomal arrays can be triggered by treating worms with X-rays, UV, or a combination of genotoxic compounds and UV. During DNA repair, arrays are occasionally

Table 2.3 The different types of *C. elegans* transgenic animals

Transgene category	Transgene location	Copy number	Transmission	Main advantage	Main limitation
Extrachromosomal array	Outside chromosomes	Up to ~200	Non-mendelian (variable rate)	Easiest, fastest, good for overexpression	Mosaicism, variable expression, constant selection needed
Integrated array	On one chromosome (random)	Up to ~200	Mendelian	Best option for mosaicism-free overexpression	Labor-intensive, side mutations may remain
Single copy integrant/ knock in	On one chromosome (chosen)	1 per haploid genome	Mendelian	Control on insertion site, genomic context maintained	No strong overexpression

integrated at a random location in the genome. The main limitations of the procedure are (i) its relatively low efficiency, such that large F_2 populations must be screened before an *integrated array line* is identified, and (ii) the generation of side-mutations, which must be removed by repeated backcrossing. Recently, a strategy using the CRISPR/Cas9 nuclease was reported to successfully trigger site-directed integration of extrachromosomal arrays [47].

Single-Copy Transgenes

Single-copy transgenes inserted in the chromosomes are particularly useful for achieving near-physiological expression levels and maintaining chromatin regulatory functions. Compared to (non-integrated) extrachromosomal array lines, the generation of single-copy transgenic animals takes more time, with several steps of animal selection and molecular verifications, but it allows for more controlled and interpretable experiments. The generation of single-copy transgenic animals involves more sophisticated approaches and relies on specific enzymatic machinery to insert transgenes via one of the three following processes: transposition, DNA cleavage/repair, or recombination. In a first method set, the Mos1 transposase is used to insert a transgene at predetermined (MosSCI) or random (miniMos) locations in the genome [48–50]. A second set of methods uses CRISPR/Cas9-induced sequence-specific DNA cleavage and subsequent homologous repair with a transgene-containing repair template [40, 51]. This versatile method allows for a wide range of genome modifications, such as point mutations, gene knockout, reporter knock-in, or the introduction of transgenes (see, e.g., [40, 52] for reviews). The efficacy of large-size transgene insertion is however relatively low. A third approach, Recombinase-Mediated Cassette Exchange (RMCE), leverages on the action of FLP recombinase enzymes to introduce transgenes into receiving strains, which have been engineered to contain compatible FRT sites at specific genomic loci [53]. Although this latter approach is very recent and has not been extensively used, it should allow efficient knock-ins of large transgenes.

2.2.2.2 How to Target the Expression of a Transgene in a Specific Neuron Type

With its 302 neurons identified at the single-cell resolution, *C. elegans* is an incredibly potent model for determining the function of single neurons within circuits controlling behavior. Many neurogenetic analyses rely on the ability to drive the expression of specific transgenes into specific neurons (see Points 2.2.4 and 2.2.5). To this end, previously characterized promoters are used to drive expression in specific neurons or neuron categories (see examples in Table 2.4). Even if recent single-cell RNAseq approaches have started to accelerate the discovery of neuron-specific promoters, their list is far from covering the entire nervous system [54, 55]. Yet, a large majority of worm neurons can be selectively targeted by implementing intersectional systems combining two promoters, as illustrated by the following two examples.

Table 2.4 Examples of neuron-type specific promoters

Promoter	Expressing neurons
gcy-8p	AFD thermosensory neuron
ttx-3p	AIY interneuron
che-2p	Ciliated sensory neurons
tph-1p	Serotonergic neurons

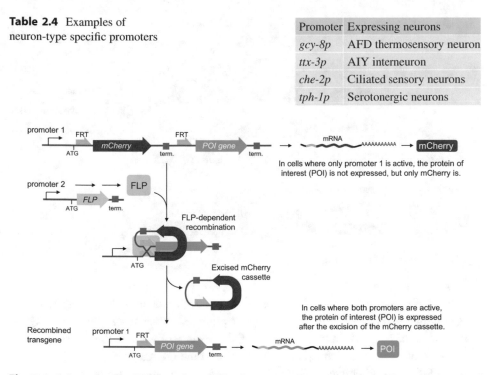

Fig. 2.6 Schematic of an FLP-based recombination system. Term: transcriptional termination signal

Recombination-Based Systems

These intersectional approaches rely on the action of a recombinase, which will modify the structure of the transgene to trigger (or sometimes abolish) transgene expression only in cells where the recombinase is expressed [56]. In the example presented in Fig. 2.6, the gene coding for a protein of interest (POI) is placed downstream of a specific promoter (promoter 1), but the POI expression is prevented by a mCherry cassette containing a Stop codon and a transcription termination signal (term.), which is placed between the promoter and the POI gene. The mCherry cassette is flanked by FRT recombination sites. A second transgene drives the expression of the FLP recombinase under the control of another promoter (promoter 2). In cells where only promoter 1 is active, mCherry is expressed, but the POI is not. In cells where promoter 2 is active, the FLP recombinase is expressed causing the mCherry cassette to be excised, such that the gene coding for the POI can now be transcribed under the control of promoter 1. Hence, the POI will only be expressed in cells where both promoters are active (intersection). A similar system using the Cre/LoxP recombinase is also available in *C. elegans* [57].

Q-system

The Q-system was derived from *Neurospora crassa* gene regulatory elements [58]. The system relies on a QUAS promoter, which is not activated by any endogenous *C. elegans*

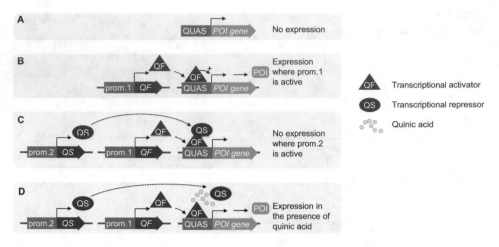

Fig. 2.7 Schematic of the Q-system

transcription factor, as well as two exogenous transcription factors: the QF activator and the QS inhibitor (Fig. 2.7a). A first *C. elegans* promoter drives the expression of QF in a specific set of neurons. In these cells, the QF protein will activate the transcription of a transgene of interest, which is under the control of the QUAS promoter (Fig. 2.7b). In addition, QS can be expressed with a different *C. elegans* promoter. Cells expressing both QF and QS will not express the QUAS-driven transgene, because QS inhibits QF (Fig. 2.7c). This permits a fraction of cells to be 'subtracted' from a list of expressing cells. In addition, the Q-system enables the temporal regulation of gene expression, since the addition of quinic acid to the growth medium prevents the inhibitory effect of QS. It is therefore possible to de-repress a given transgene at a given timepoint in cells expressing both QF and QS (Fig. 2.7d).

2.2.3 Gene Expression Analysis in the Nervous System

Researchers often need to determine where a given gene is expressed in *C. elegans* and identify the neurons that express it. This could, for example, be the immediate follow-up after a genetic screen and the identification of relevant mutations in uncharacterized genes [59].

2.2.3.1 Gene Expression Analysis with Reporter

The most conventional approach to assess the neuronal expression pattern of candidate genes is to use fluorescent protein reporters, which can be easily detected *in vivo* by fluorescence microscopy through the animal transparent body. Other gene expression analysis methods, such as fluorescence *in situ* hybridization [60] or immunocytochemistry [61], are possible in *C. elegans*, but are more challenging on a technical standpoint and

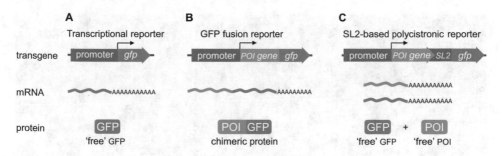

Fig. 2.8 Schematic of different types of reporter transgenes and their products

have been less extensively used, in particular for neurobiological research. The rest of this section will focus on fluorescent-reporter methods.

Types of Reporters

Fluorescent reporter transgenes [59] combine the coding sequence of a fluorescent protein with some endogenous regulatory sequences. We will consider here three examples: transcriptional reporters, translational fusion reporters, and polycistronic reporters (Fig. 2.8). In **transcriptional reporters**, the endogenous sequence typically includes only 5′ regulatory regions (usually referred to as *promoters*), resulting in the expression of a 'free' fluorescent reporter protein that reports the level of transcriptional initiation (Fig. 2.8a). It is also possible to include a gene-specific 3′UTR to report its regulatory impact (not shown). In **translational fusion reporters** (e.g., GFP fusions), the fluorescent protein gene is fused in frame with the coding sequence of the gene of interest, producing a chimeric fluorescent protein (Fig. 2.8b). Transgenes usually also encompass candidate regulatory regions, which may include intragenic elements (such as intronic elements if a genomic fragment is used). Translational fusion reporters have the potential to report both transcriptional and post-transcriptional regulation, as well as the subcellular localization of the resulting protein. In **polycistronic reporters**, the fluorescent protein sequence is separated from the endogenous sequence by a Splice Leader 2 (SL2) sequence [62] (Fig. 2.8c). This element triggers a trans-splicing event, which results in the generation of two distinct mRNAs: one coding for the endogenous gene product, the other coding for the reporter protein. In this case, the reporter protein will not inform on the subcellular localization of the protein under study, but the polycistronic approach get rids of other caveats associated with chimeric proteins. For translational reporters and polycistronic reporters, overexpression may cause unpredictable biases and care should be taken in interpreting the reporter signal. Using these constructs to rescue a mutant phenotype may provide indications for functionally relevant expression patterns/levels.

Identifying Neurons

A particularly challenging aspect of gene expression analysis in the *C. elegans* nervous system is to determine the identity of the expressing neurons. To this end, researchers

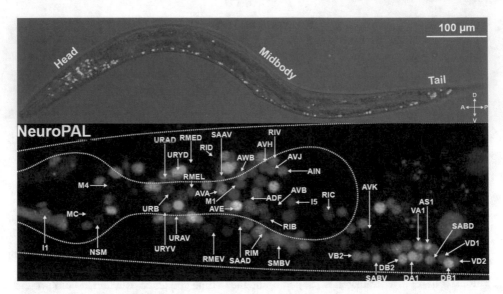

Fig. 2.9 Neuronal Polychromatic Atlas of Landmarks (NeuroPAL). Picture of an adult transgenic NeuroPAL worm overlaying a bright-field micrograph with multi-channel fluorescence signals (top panel). Close-up view of the fluorescence signal in the animal head with dotted lines highlighting the pharynx and the animal contour (bottom panel). Reproduced from the NeuroPAL manual with authorization (courtesy of Eviatar Yemini)

typically need to consider several criteria, including (i) the morphology and the number of neurites (unfortunately of limited discriminatory power), (ii) the morphology of sensory endings (limited to amphid sensory neurons and requiring a strong reporter expression), and (iii) the neuron cell body position, preferably in comparison to the position of other identified neurons. Such landmark neurons can be highlighted by treating the animal with DiL (red) or DiO (green) dyes, which fill a defined subset of sensory neurons exposed to the environment. Alternatively, transgenic markers can be used, which requires co-injecting known markers or using a specific 'pre-colored' transgenic background with labeled landmark neurons [63]. In the last few years, Yemini and collaborators have developed a "brainbow-like" transgenic background that assembles multiple fluorescent reporters to create predetermined combinations in most animal neurons (Fig. 2.9). The so-called Neuronal Polychromatic Atlas of Landmarks (NeuroPAL) can be used to significantly streamline the identification of neurons [64].

2.2.3.2 Neuron-specific Transcriptomics

Beyond reporter methods analyzing a single gene at a time, *C. elegans* is also amenable to large-scale tissue/cell-specific transcriptomics. The first neuron type-specific transcriptomic data were obtained after isolation of neuronal precursors from embryos and differentiation *in vitro* [65]. Over the years, several methods have emerged to circumvent the inherent challenges caused by the tough worm cuticle and the difficulties in isolating

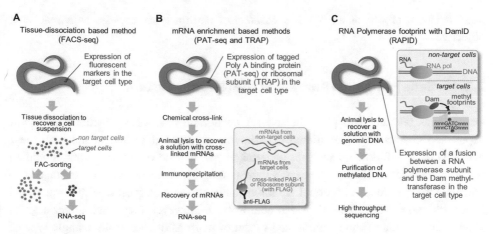

Fig. 2.10 Methods for neuron-specific transcriptomic analysis

specific *C. elegans* cell types. Three main strategies have been developed to analyze neuron-type-specific transcription *in vivo*. All rely on the ability to express transgenes in the targeted cells, and therefore on an expression system established *a priori*. A first approach involves the dissociation of animals, FAC-sorting of fluorescently labeled cells, and RNAseq analyses [66] (Fig. 2.10a). A second approach set relies on the isolation of tagged RNA via the immunoprecipitation of cross-linked flagged Poly-A-binding proteins (FLAG::PAB-1) or cross-linked ribosomal subunits expressed in a specific cell type [67, 68] (Fig. 2.10b). All these approaches remain technically challenging to execute and are usually not routinely applied in most laboratories. Recently, a cell-specific RNA polymerase footprinting methods, based on the *D*NA *a*denosine *m*ethyltransferase *id*entification (DamID) technology [69], was introduced and could represent a sensitive and cost-effective alternative to identify genes transcribed in specific neuron types (Fig. 2.10c).

2.2.3.3 Single-Cell RNA-seq to Learn About Every Cell Type in a Single Analysis

Recent technological developments in single-cell RNA sequencing (scRNA-seq) have opened new doors to start cataloging more comprehensively the transcriptional profile of individual neuron types in *C. elegans*. Applied to L2 larvae, this approach has already demonstrated the possibility to identify sets of transcripts expressed in different neuron types [55, 70]. The CeNEGEN consortium has developed a pipeline combining FAC-sorting of neuron classes with scRNA-seq [71]. Initial results are extremely encouraging, showing that clusters corresponding to most neuron types can be separated from L4 larvae [54]. Although the transcriptome coverage is still limited (many cell clusters gathered only a few dozen cells and had only a few hundred transcripts detected), there are no inherent limitations preventing to upscale the approach. Ultimately, obtaining a comprehensive transcriptional profile of all the 118 *C. elegans* neuron types appears as a reachable objective.

2.2.4 Gene Functional Analysis in the Nervous System

Once genes relevant to a given behavioral phenotype have been identified and their expression characterized, additional experiments are needed to understand how these genes work. The follow-up analyses undertaken will depend on many factors, including previous knowledge of these genes and their homologs. A major goal at this stage is to elucidate the gene locus of action. Like for virtually any animal, most *C. elegans* neuronal genes are expressed in more than one neuron type. However, they often affect a given phenotype by acting from a determined subset of neurons or even from a single type of neurons. Therefore, identifying the locus of action of a given gene is often a key step in elucidating the causes of a mutant phenotype, as it both delineates the candidate neural circuit and orients further experiments to particular neuron types. Two main "generic" approaches, potentially applicable to any gene of interest, can be used to determine the locus of action of neuronal genes.

First, *neuron-specific rescue* experiments in mutants can provide information about the neurons in which the gene (re)-expression is sufficient to recover a wild-type phenotype (see Fig. 2.11). Extrachromosomal array lines are often used for rescue experiments when

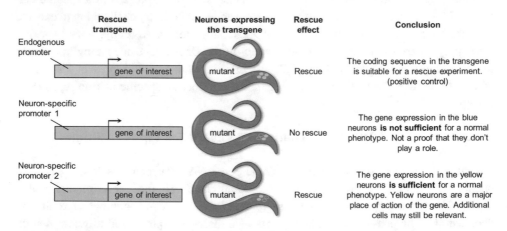

Fig. 2.11 Principle of a neuron-specific rescue approach. When a loss-of-function mutation causes an abnormal phenotype, neuron-specific rescue can be performed to identify the neuronal locus of action of the mutated gene. Experiments are carried out in this loss-of-function mutant. In a positive control experiment (upper), the endogenous promoter of the gene is used to restore expression in all neurons where the gene is normally expressed (depicted in blue and yellow). In neuron-specific rescue conditions (middle and lower), neuron-specific promoters are used to restore gene expression in selected target neurons (either blue neurons or yellow neurons)

several candidate neurons need to be tested with a battery of promoters. Although an aberrant overexpression effect cannot be totally excluded, a positive rescue effect is a solid indication for a major site of action of the gene. In contrast, a negative result should not be taken as a proof that the gene has no role in a given neuron type. Indeed, it is possible that the gene plays a role in that cell type, but that the parallel expression in other neurons is also required to produce a detectable rescue effect. Furthermore, a lack of rescue could be explained by transgene mosaicism, the inability of the rescue construct to yield proper isoform expression, or its failure to recapitulate quantitative and/or temporal aspects of the endogenous gene expression.

Second, *neuron-specific knockdown* of a given gene can provide information about the neuron(s) in which a given gene is required to avoid an abnormal phenotype. Neuron-specific knockdown can be achieved by co-expressing sense and antisense RNA stretches that are complementary to the targeted gene product [72]. The double-stranded RNA (dsRNA) will trigger an RNA*i* effect and cause the knockdown of the targeted mRNA, but only in dsRNA-expressing cells (Fig. 2.12). Alternative approaches include the expression of dominant negative mutants (when available) and neuron-specific gene knockout. In the latter case, genome editing can be used to introduce loxP or FRT sites around the target

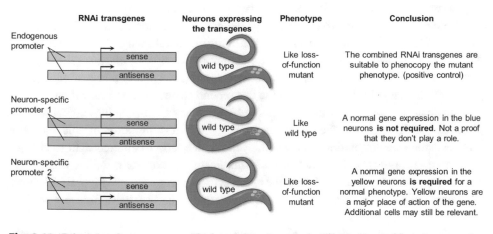

Fig. 2.12 Principle of a neuron-specific knockdown approach. When a loss-of-function mutation causes an abnormal phenotype, neuron-specific RNAi can be triggered to identify the neuronal locus of action of the mutated gene. Experiments are carried out in a wild-type background with the aim of causing a phenotype similar to that observed in mutants (phenocopy). In a positive control experiment (upper), the endogenous promoter of the gene is used to express sense and antisense RNAs, which will form double-stranded RNA and knock down the gene of interest via RNAi in all neurons where the gene is normally expressed (depicted in blue and yellow). In neuron-specific knockdown experiments (middle and lower), neuron-specific promoters are used to trigger RNAi in selected target neurons (either blue neurons or yellow neurons). Additional controls (not shown) should include animals with only the sense and antisense transgenes, respectively. Note that the same framework can be applied when using alternative methods for gene knockdown, such as when expressing dominant negative mutant genes or using a recombination-based cell-specific knockout

gene and neuron-specific knockout can be triggered by a transgene driving the expression Cre or FLP recombinases in target neurons [73, 74].

2.2.5 Circuit Functional Analysis

A comprehensive understanding of behavior and its link to underlying neural circuit function is possible in *C. elegans*. This subsection presents some of the experimental methods used to study neural circuit functioning. These methods rely on the generation of transgenic animals, on the possibility to drive the expression of transgenes in specific neurons (see Point 2.2.2), and on the availability of tools to monitor, upregulate, or downregulate neuronal activity (see below).

2.2.5.1 Monitoring Neuronal Activity

Neuronal activation is concomitant with an increase in intracellular calcium ion concentration. Genetically-encoded calcium indicators have been widely used to monitor neural activity in cultured cells and tissues [75]. Thanks to its transparent body and small size, *C. elegans* can be used to monitor neuronal activity *in vivo*. Two types of indicators have been mostly used in worms: GCaMP/RCaMP and Cameleon.

GCaMP is an artificial protein that fuses the calcium-binding domain of Calmodulin and GFP [76, 77]. Calcium binding results in a conformational change that enhances green fluorescence. RCaMP, a red version, is also commonly used in *C. elegans* [78]. In order to correct for motion artifacts, it is advisable to record the signal of a second fluorescent protein of another color in parallel and use this second channel for normalization.

Cameleons are also artificial proteins combining fluorescent proteins and calmodulin [79]. The FRET (fluorescence resonance energy transfer) property of this indicator changes upon calcium binding, causing the yellow/blue ratio of emitted light to change. Ratiometric measures with Cameleons will implicate the excitation with a single light wavelength and emission recording in the two specific emission ranges. Figure 2.13 shows an example of heat-evoked calcium responses recorded in the FLP thermosensory neurons [80].

By developing and implementing sophisticated microscopy equipment, a few laboratories have started to record brain-wide activity (i.e., monitoring large numbers of neurons in the same experiments) in immobilized animals, or even in freely moving animals [81–84]. While one of the current bottlenecks in this type of analysis remains the identification of the recorded neurons, the recent implementation of NeuroPAL represents a major advance toward an exhaustive monitoring of the worm brain activity at the identified-neuron resolution [64].

2.2.5.2 Upregulating Neuronal Activity

Being able to upregulate the activity of selected neurons and quantify the behavioral outcome is a very useful strategy for inferring the function of specific building blocks

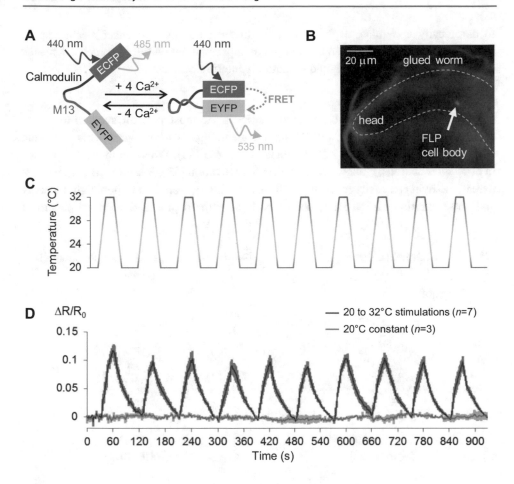

Fig. 2.13 Principle and example of Cameleon-based analysis of neuronal activity (**a**) Principle of the Cameleon indicator. (**b**) A glued worm with fluorescence signal in the cell body of FLP neurons (only CFP signal is shown). (**c**) Heat stimulation protocol. (**d**) CFP/YFP ratio increase over baseline (DR) recorded in stimulated FLP cell bodies (red trace) or in control experiments at constant temperature (blue trace). Mean (solid trace) and s.e.m. (colored shade). Values are normalized by dividing by the ratio during the baseline (R_0).

of the *C. elegans* nervous system. We will present here two types of 'manipulations' able to up-regulate the activity of targeted neurons, either permanently or under tight temporal control.

Enhancement of Neurotransmission

A simple method to enhance the synaptic output of targeted neurons involves the neuron-specific expression of a PKC-1 gain-of-function mutant [85, 86]. Because PKC-1 activity is known to enhance neurotransmission, this approach is interesting when one wants to chronically upregulate a given neuron. However, because the exact mechanism of PKC-dependent overactivity and its long-term impact cannot be fully appreciated, results obtained with this approach should be carefully interpreted.

Optogenetic Cell Activation

Channelrhodopsin-2 (ChR2)-based optogenetics is the most versatile method to hyper-activate (depolarize) neurons. ChR2 from the green algae *Chlamydomonas* is a cationic channel that opens upon blue light illumination (Fig. 2.14a). When ectopically expressed in *C. elegans* neurons, it allows their on-demand activation [87]. Because worms are transparent, the light can easily reach any ChR2-expressing neuron. In addition, light intensity and pulse duration can be easily controlled to vary the degree and temporal pattern of

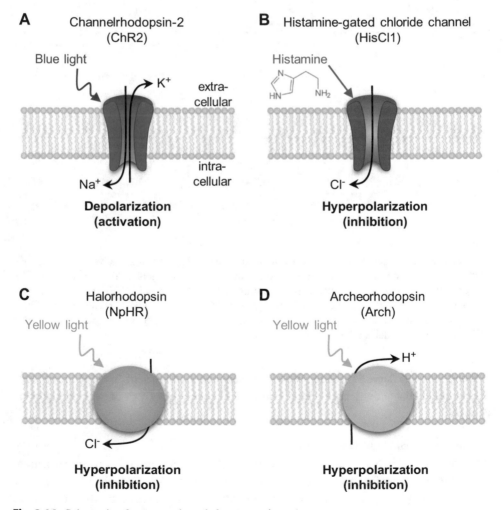

Fig. 2.14 Schematic of optogenetic and chemogenetic tools

activation. It should be noted that worms display some endogenous responsiveness to intense blue lights. However, the caveat related to endogenous blue-light sensitivity can be circumvented by using a light-insensitive mutant background (*lite-1*), more sensitive Channelrhodopsin variants (such as CoChR, which requires low light intensities), or red light-sensitive variants (such as Chrimson) [88–90]. Optogenetics methods are incredibly potent to upregulate neural activity, determine the phenotypic (behavioral) impact of a narrow neural pathway, or establish the polarity of synapses (excitatory versus inhibitory). However, it is important to remember that, in most instances, experimenters do not know to what extent light-evoked cellular activity will recapitulate a normal neural activity.

2.2.5.3 Downregulating Neuron Activity

The downregulation of specific neurons is a relatively straightforward approach to infer their function. We will review here some genetic approaches used in *C. elegans*, which involve different inhibition modes, which are more or less invasive, and which are either irreversible or temporally controlled. In order to interpret results, it is important to keep in mind that the neural circuit may rewire, either structurally or functionally, in response to these manipulations. This occurs especially when the changes are permanent and probably even more if they occur during animal development.

Neuron Killing

In the laboratory, most neurons are dispensable for worm growth (in particular the sensory neurons). Laser-ablation of fluorescently labeled neurons was the first method used to infer their function. This approach was used, for example, to demonstrate that touch receptor neurons are indeed required for gentle touch-evoked responses. While laser-ablation is still used nowadays, it requires specific equipment and the generation of ablated animals is low throughput. As complementary approaches, several genetic ablation methods have been developed to kill neurons. A first approach consists in expressing a gain-of-function degenerin channel (MEC-4) mutant, which is constitutively open and which can cause neuronal death [91]. A second approach consists in expressing caspases. Caspases are late effectors in the control of programmed cell death (apoptosis). Human caspase-3 or worm CED-3 proteins are activated after their cleavage into two peptides that form heterotetramers to trigger death signals. A reconstituted caspase system has been used to ablate specific neurons [92]. The two peptides can be expressed from different promoters, and will only induce apoptosis where both subunits are expressed, providing an intersectional system to spatially restrict cell ablation. A third approach consists in producing oxidative damage to destroy cells. MiniSOGs are small proteins that catalyze the synthesis of destructive singlet oxygen when expressed in mitochondria and exposed to light [93]. MiniSOGs can be selectively expressed in target neurons and activated by illuminating the animals at a chosen time point in order to accurately define the onset of cell destruction.

Inhibition of Neurotransmission

Tetanus Toxin (TeTx) inhibits synaptic transmission by cleaving the synaptic vesicle protein synaptobrevin. Synaptobrevin is one of the proteins involved in the formation of the SNARE complexes that drive membrane fusion during exocytosis. TeTx expression can cause a functional inhibition when transgenically expressed in worms [94]. In principle, this type of inhibition will not cause cell death, nor will it affect communication through Gap junctions. As such, it is thus less invasive than the neuron ablation methods presented above. However, unless a temporally-controlled TeTx expression system is used, the permanent inhibition of neurotransmission may still cause some circuit functional rewiring.

Chemogenetic Cell Inhibition

Histamine is apparently not used as a neurotransmitter by *C. elegans*. Histamine-gated chloride channels can be expressed in specific *C. elegans* neurons [95]. In the presence of histamine, the opening of the chloride channel will cause neuronal hyperpolarization and decreased activity (Fig. 2.14b). This chemogenetic approach is well-suited for experiments lasting from a few minutes to a few tens of minutes. Indeed, effects over shorter durations are limited by the time needed for histamine diffusion, whereas effects over longer durations may be precluded by intrinsic homoeostatic mechanisms able to compensate for the artificial hyperpolarization.

Optogenetic Cell Inhibition

Optogenetic interventions are usually well-suited to induce short-term (seconds to minutes) responses. Several optogenetic tools have been adapted to *C. elegans* to reversibly inhibit neuronal activation in a temporally-controlled manner [see 96 for a review]. These include Halorhodopsin from *Natronomonas* (NpHR), anion channelrhodopsins (ACRs) from the alga *Guillardia theta* [97] and proton pumps, such as Archaerhodopsin-3 (Arch). The activation of NpHR or ACRs leads to an influx of chloride ions and the activation of proton pumps leads to proton ex-flux, both causing neurons to hyperpolarize, thereby reducing their excitability (Fig. 2.14c and d).

2.3　Perspectives on Nervous System Function Analyses in *C. elegans*

Over the past few decades, the possibilities for monitoring and controlling neuronal activation have substantially grown. There is still room for considerable improvements, but progresses are constantly being made and these tools are becoming more and more efficient. Combined with the relative simplicity of *C. elegans* nervous system, the availability of a full synaptic connectivity map, and several methods for quantifying behavior, modern neurogenetics methods hold the promise to enable a very thorough understanding of how the *C. elegans* nervous system controls behavior. Anticipated progress of neurogenetics in the future can be highlighted in the following key areas:

2.3.1   Application of State-of-the-art Methods to Better Understand the Function of Gene, Neuron, Circuit, and Behavior

Recently developed cutting-edge tools to study gene, neuron, circuit, and behavior have shown enormous potential to broaden the horizon of understanding nervous system function in *C. elegans*. We can apply newly developed single-cell transcriptomics (CeNGEN project) or neuron-specific transcriptomic methods (TRAP, RAPID) to study the transcriptional program in all the neurons or specific neurons of interest. Multi-scale neuronal activation, inhibition and imaging tools can reveal neuronal dynamics and its function in the behavior from seconds to minutes timescales. Whole-animal calcium imaging methods combined with neuron identification tool (NeuroPAL) can reveal the circuit dynamics in response to a single sensory stimulus, multiple stimuli, during a specific behavior or during global behavioral state changes. Finally, high-content behavioral tracking can unravel changes in behavioral micro-components such as reversal, speed, and posture of worms. Progress in our understanding of the nervous system will depend on the expanding application of these tools to answer questions relevant for neuronal plasticity, health, and disease. This will require continuous efforts to bring the costs down and make these tools accessible for many labs.

2.3.2   Integration of Multiple Methods to Bridge the Function of Genes, Neurons, and Circuits to Behavior

One of the key challenges in modern neuroscience is to provide a holistic understanding of the nervous system. Newly developed neurogenetic tools can advance our understanding of a single aspect (gene or neuron or circuit or behavior) of the nervous system function, but usually cannot directly help us to bridge the gap across these different levels. Integrative approaches, studying multiple aspects through different methods in the same study, are required and this is a strategy for which *C. elegans* is particularly well suited (Fig. 2.4). Nevertheless, performing such studies remains difficult due to three main reasons: (i) unavailability of multiple methods in the same lab, (ii) considerable time requirement to complete the study, and (iii) technical challenges involved in integrating different methods and events sometimes occurring on different timescales. For example, understanding how a neural circuit functions to control a behavior of interest would require us to integrate worm tracking methods with neuron function manipulation while concomitantly imaging the entire nervous system activity. This task is technically challenging. Furthermore, a definitive understanding of the functioning of such a complex system will ultimately require scientists to be able to simulate it and to create accurate predictions. Along this line, a series of ambitious software development projects are articulated by the OpenWorm Foundation, which aims at creating an integrative simulation of *C. elegans* in order to understand how the animal behavior may arise from its fundamental biological components [98].

2.3.3 Development of Cost-Effective Novel Neurogenetic Tools

The future of our in-depth understanding of the nervous system depends on the development of novel cost-effective neurogenetic methods. This critical requirement has led to the emergence of the new field of synthetic neurobiology. The main objective of synthetic neurobiology is to develop new tools for DNA, RNA, or protein-based bio-engineering and improve robotics, electronics, and imaging methods for neuroscience. The simple nervous system of *C. elegans* provides an excellent platform to test *in vivo* applications of these tools. Integration of these methods can extensively boost our understanding of genes, neurons, circuit, and behavior.

2.3.3.1 Gene Function

To better understand the function of the genes involved in a physiological process, we need tools to monitor and modulate gene or protein activity with high spatial and temporal resolution. Monitoring protein activity is now possible with fluorescence-based biosensors [99–101]. These sensors can be designed for a protein of interest through protein engineered based on the cellular function of that protein. To modulate genes expression in the cell, we can combine protein engineering with non-invasive methods as optogenetics [102–104]. One of the key challenges here will be to make these tools functional *in vivo* at single-neuron resolution in *C. elegans*. Development and *in vivo* integration of these tools for more genes would allow for a better understanding of cell signaling involved in various physiological processes.

2.3.3.2 Neuron Function

C. elegans neuroscience has made enormous progress in the last decade to highlight the role of single neurons and circuits in the emergence of specific behaviors, thanks to optogenetic methods for neuronal activity monitoring (GCaMP, Cameleon) and stimulation (Channelrhodopsin). Now, the development of novel neurogenetic tools is required to gain insights into altered neuronal physiology after plasticity, aging, and diseases. We need to probe the impact of these processes on the function of various organelles, inter-organelle signaling, and dendritic/synaptic physiology and morphology. Here, a major limitation will relate to the small size of *C. elegans* neurons.

2.3.3.3 Circuit Architecture and Function

Circuits underlying behaviors can be identified by combining methods to ablate neurons and block neuronal activity. As highlighted above, we can gain a deeper understanding of circuit dynamics by integrating whole animal imaging methods with neuron stimulation via optogenetics. This method is particularly well suited to decode feedforward and feedback loops in neural pathways underlying behaviors. Development of slow kinetics activity reporter (minutes timescale) compared to fast calcium sensors (seconds timescales) could potentially help in the identification of entire recruited circuits after a behavioral experiment is finished and solve the complications owing to the need for simultaneously monitoring and stimulating neurons with light during optogenetic-based experiments. In addition, recently highlighted structural plasticity in the worm connectome [14] calls for the expansion of optical microscopy tools able to monitor the appearance and disappearance of synapses *in vivo*.

2.3.4 Translational Prospects

Continuous advancement in *C. elegans* neurogenetics holds enormous promises for academic labs and pharmaceutical industry. In the upcoming decade, these advancements could be implemented to decode fundamental physiological and pathophysiological mechanisms. Using *C. elegans* for high-throughput screening of genetic pathways implicated in neurological diseases and for potential drugs (including in transgenic humanized worms expressing human proteins) can accelerate advances in the discovery of new therapeutic pathways. Combining the simplicity of *C. elegans* with advanced tools, the future of basic and translational neurogenetic research holds a bright spot.

> **Take-Home message**
> *C. elegans* is a potent neurogenetics model to obtain a comprehensive understanding of the nervous system structure and function at the gene, neuron, circuit, and behavior levels and to carry out integrative experiments needed to understand the connection between the processes taking place across these different levels and over different timescales.

Acknowledgments This work was supported by the Swiss National Science Foundation (SNSF, grants IZCNZ0_174703/SBFI_C16.0013, BSSGI0_155764, PP00P3_150681, and 310030_197607 to D.A.G).

References

(Owing to space limitations, the authors apologize that many articles that significantly contributed to *C. elegans* neurogenetics could not be cited.)

1. Nigon VM, Felix M-A. History of research on *C. elegans* and other free-living nematodes as model organisms, in WormBook, T.C.e.R. Community, Editor. 2017, WormBook.
2. Ankeny RA. The natural history of *Caenorhabditis elegans* research. Nat Rev Genet. 2001;2(6):474–9.
3. Consortium, T.C.e.S., Genome sequence of the nematode *C. elegans*: a platform for investigating biology. Science. 1998:282(5396);2012.
4. Corsi AK, Wightman B, Chalfie M. A transparent window into biology: a primer on *Caenorhabditis elegans*. Genetics. 2015;200(2):387–407.
5. Kaletta T, Hengartner MO. Finding function in novel targets: *C. elegans* as a model organism. Nat Rev Drug Discov. 2006;5(5):387–98.
6. Culetto E, Sattelle DB. A role for *Caenorhabditis elegans* in understanding the function and interactions of human disease genes. Hum Mol Genet. 2000;9(6):869–77.
7. Hobert O. The neuronal genome of *Caenorhabditis elegans*, in WormBook, T.C.e.R. Community, Editor. 2013, WormBook.
8. Liu Q, et al. *C. elegans* AWA olfactory neurons fire calcium-mediated all-or-none action potentials. Cell. 2018;175(1):57–70.e17.
9. Li C, Kim K. Neuropeptides. WormBook, 2008:1–36.
10. Altun ZFH, Herndon LA, Wolkow CA, Crocker C. Lints R, Hall DH. WormAtlas. 2002–2021. Available from: http://www.wormatlas.org.
11. Pierce-Shimomura JT, et al. The homeobox gene lim-6 is required for distinct chemosensory representations in *C. elegans*. Nature. 2001;410(6829):694–8.
12. White JG. Getting into the mind of a worm – a personal view, in WormBook, T.C.e.R. Community, Editor. 2013, WormBook.
13. Cook SJ, et al. Whole-animal connectomes of both *Caenorhabditis elegans* sexes. Nature. 2019;571(7763):63–71.
14. Witvliet D, et al. Connectomes across development reveal principles of brain maturation in *C. elegans*. bioRxiv. 2020: p. 2020.04.30.066209.
15. de Bono M, Maricq AV. Neuronal substrates of complex behaviors in *C. elegans*. Annu Rev Neurosci. 2005;28:451–501.
16. Schafer WR. Deciphering the neural and molecular mechanisms of *C. elegans* behavior. Curr Biol. 2005;15(17):R723–9.
17. Faumont S, Lindsay TH, Lockery SR. Neuronal microcircuits for decision making in *C. elegans*. Curr Opin Neurobiol. 2012;22(4):580–91.
18. Liu P, Chen B, Wang Z-W. GABAergic motor neurons bias locomotor decision-making in *C. elegans*. Nat Commun. 2020;11(1):5076.
19. Cohen D, et al. Bounded rationality in *C. elegans* is explained by circuit-specific normalization in chemosensory pathways. Nat Commun. 2019;10(1):3692.
20. Barrios A. Exploratory decisions of the *Caenorhabditis elegans* male: a conflict of two drives. Semin Cell Dev Biol. 2014;33:10–7.
21. Hart AC (ed.). Behavior, in WormBook, T.C.e.R. Community, Editor. 2006, WormBook.
22. Mahoney TR, Luo S, Nonet ML. Analysis of synaptic transmission in *Caenorhabditis elegans* using an aldicarb-sensitivity assay. Nat Protoc. 2006;1(4):1772–7.
23. Husson SJ, et al. Keeping track of worm trackers. WormBook. 2013:1–17.

24. Javer A, et al. An open-source platform for analyzing and sharing worm-behavior data. Nat Methods. 2018;15(9):645–6.
25. Ahringer J. (ed.). Reverse genetics, in WormBook, T.C.e.R. Community, Editor. 2006, WormBook.
26. Brenner S. The genetics of *Caenorhabditis elegans*. Genetics. 1974;77(1):71–94.
27. Jorgensen EM, Mango SE. The art and design of genetic screens: *Caenorhabditis elegans*. Nat Rev Genet. 2002;3(5):356–69.
28. Kutscher LM, Shaham S. Forward and reverse mutagenesis in *C. elegans*, in WormBook, T.C.e.R. Community, Editor. 2014, WormBook.
29. Doitsidou M, et al. *C. elegans* mutant identification with a one-step whole-genome-sequencing and SNP mapping strategy. PLoS One. 2010;5(11):e15435.
30. Fay DS. Classical genetic methods, in WormBook, T.C.e.R. Community, Editor. 2013, WormBook.
31. Doitsidou M, Jarriault S, Poole RJ. Next-generation sequencing-based approaches for mutation mapping and identification in *Caenorhabditis elegans*. Genetics. 2016;204(2):451–74.
32. Chalfie M, Au M. Genetic control of differentiation of the *Caenorhabditis elegans* touch receptor neurons. Science. 1989;243(4894):1027.
33. Ernstrom GG, Chalfie M. Genetics of sensory mechanotransduction. Annu Rev Genet. 2002;36:411–53.
34. Fire A, et al. Potent and specific genetic interference by double-stranded RNA in *Caenorhabditis elegans*. Nature. 1998;391(6669):806–11.
35. Conte D. Jr, et al. RNA interference in *Caenorhabditis elegans*. Curr Prot Mol Biol. 2015. 109:26.3.1–26.3.30.
36. Kamath RS, et al. Systematic functional analysis of the *Caenorhabditis elegans* genome using RNAi. Nature. 2003;421(6920):231–7.
37. Boutros M, Ahringer J. The art and design of genetic screens: RNA interference. Nat Rev Genet. 2008;9(7):554–66.
38. Calixto A, et al. Enhanced neuronal RNAi in *C. elegans* using SID-1. Nat Methods. 2010;7(7):554–9.
39. Evans TC (ed.). Transformation and microinjection. WormBook, 2006.
40. Nance J, Frøkjær-Jensen C. The *Caenorhabditis elegans* transgenic toolbox. Genetics. 2019;212(4):959.
41. Mello CC, et al. Efficient gene transfer in *C.elegans*: extrachromosomal maintenance and integration of transforming sequences. EMBO J. 1991;10(12):3959–70.
42. Stinchcomb DT, et al. Extrachromosomal DNA transformation of *Caenorhabditis elegans*. Mol Cell Biol. 1985;5(12):3484–96.
43. Granato M, Schnabel H, Schnabel R. pha-1, a selectable marker for gene transfer in *C. elegans*. Nucleic Acids Res. 1994;22(9):1762–3.
44. Miyabayashi T, et al. Expression and function of members of a divergent nuclear receptor family in *Caenorhabditis elegans*. Dev Biol. 1999;215(2):314–31.
45. Praitis V, et al. Creation of low-copy integrated transgenic lines in *Caenorhabditis elegans*. Genetics. 2001;157(3):1217–26.
46. Radman I, Greiss S, Chin JW. Efficient and rapid *C. elegans* transgenesis by bombardment and hygromycin B selection. PLoS One. 2013;8(10):e76019.
47. Yoshina S, et al. Locus-specific integration of extrachromosomal transgenes in *C. elegans* with the CRISPR/Cas9 system. Biochem Biophys Reports. 2016;5:70–6.
48. Frokjaer-Jensen C, et al. Improved Mos1-mediated transgenesis in *C. elegans*. Nat Methods. 2012;9(2):117–8.

49. Frokjaer-Jensen C, et al. Single-copy insertion of transgenes in *Caenorhabditis elegans*. Nat Genet. 2008;40(11):1375–83.
50. Frøkjær-Jensen C, et al. Random and targeted transgene insertion in *Caenorhabditis elegans* using a modified Mos1 transposon. Nat Methods. 2014;11(5):529–34.
51. Frokjaer-Jensen C. Exciting prospects for precise engineering of *Caenorhabditis elegans* genomes with CRISPR/Cas9. Genetics. 2013;195(3):635–42.
52. Dickinson DJ, Goldstein B. CRISPR-based methods for *Caenorhabditis elegans* genome engineering. Genetics. 2016;202(3):885–901.
53. Nonct ML. Efficient transgenesis in *Caenorhabditis elegans* using flp recombinase-mediated cassette exchange. Genetics. 2020;215(4):903.
54. Taylor SR, et al. Expression profiling of the mature *C. elegans* nervous system by single-cell RNA-Sequencing. bioRxiv, 2019:737577.
55. Lorenzo R, et al. Combining single-cell RNA-sequencing with a molecular atlas unveils new markers for *Caenorhabditis elegans* neuron classes. Nucleic Acids Res. 2020;48(13):7119–34.
56. Davis MW, et al. Gene activation using FLP recombinase in *C. elegans*. PLoS Genet. 2008;4(3):e1000028.
57. Hoier EF, et al. The *Caenorhabditis elegans* APC-related gene apr-1 is required for epithelial cell migration and Hox gene expression. Genes Dev. 2000;14(7):874–86.
58. Wei X, et al. Controlling gene expression with the Q repressible binary expression system in *Caenorhabditis elegans*. Nat Methods. 2012;9(4):391–5.
59. Boulin T, Etchberger JF, Hobert O. Reporter gene fusions, in WormBook, T.C.e.R. Community, Editor. 2006, WormBook.
60. Bolková J, Lanctôt C. Quantitative gene expression analysis in *Caenorhabditis elegans* using single molecule RNA FISH. Methods. 2016;98:42–9.
61. Duerr JS. Immunostainings in nervous system development of the nematode *C. elegans*. Methods Mol Biol. 2020;2047:293–310.
62. Spieth J, et al. Operons in *C. elegans*: polycistronic mRNA precursors are processed by trans-splicing of SL2 to downstream coding regions. Cell. 1993;73(3):521–32.
63. Toyoshima Y, et al. Neuron ID dataset facilitates neuronal annotation for whole-brain activity imaging of *C. elegans*. BMC Biol. 2020;18(1):30.
64. Yemini E, et al. NeuroPAL: a multicolor atlas for whole-brain neuronal identification in *C. elegans*. Cell. 2021;184(1):272–288.e11.
65. Colosimo ME, et al. Identification of thermosensory and olfactory neuron-specific genes via expression profiling of single neuron types. Curr Biol. 2004;14(24):2245–51.
66. Kaletsky R, et al. The *C. elegans* adult neuronal IIS/FOXO transcriptome reveals adult phenotype regulators. Nature. 2016;529(7584):92–6.
67. Von Stetina SE, et al. Cell-specific microarray profiling experiments reveal a comprehensive picture of gene expression in the *C. elegans* nervous system. Genome Biol. 2007;8(7):R135.
68. Rhoades JL, et al. ASICs mediate food responses in an enteric serotonergic neuron that controls foraging behaviors. Cell. 2019;176(1):85–97.e14.
69. Gómez-Saldivar G, et al. Tissue-specific transcription footprinting using RNA Pol DamID (RAPID) in *Caenorhabditis elegans*. Genetics. 2020;216(4):931–45.
70. Cao J, et al. Comprehensive single-cell transcriptional profiling of a multicellular organism. Science. 2017;357(6352):661–7.
71. Hammarlund M, et al. The CeNGEN project: the complete gene expression map of an entire nervous system. Neuron. 2018;99(3):430–3.
72. Esposito G, et al. Efficient and cell specific knock-down of gene function in targeted *C. elegans* neurons. Gene. 2007;395(1–2):170–6.

73. López-Cruz A, et al. Parallel multimodal circuits control an innate foraging behavior. Neuron. 2019;102(2):407–419.e8.
74. Harterink M, et al. Local microtubule organization promotes cargo transport in *C. elegans* dendrites. J Cell Sci. 2018;131(20):jcs223107.
75. Kerr RA. Imaging the activity of neurons and muscles, in WormBook, T.C.e.R. Community, Editor. 2006, WormBook.
76. Tian L, et al. Imaging neural activity in worms, flies and mice with improved GCaMP calcium indicators. Nat Methods. 2009;6(12):875–81.
77. Nakai J, Ohkura M, Imoto K. A high signal-to-noise Ca(2+) probe composed of a single green fluorescent protein. Nat Biotechnol. 2001;19(2):137–41.
78. Akerboom J, et al. Genetically encoded calcium indicators for multi-color neural activity imaging and combination with optogenetics. Front Mol Neurosci. 2013;6
79. Miyawaki A, et al. Fluorescent indicators for Ca2+ based on green fluorescent proteins and calmodulin. Nature. 1997;388(6645):882–7.
80. Saro G, et al. Specific Ion channels control sensory gain, sensitivity, and kinetics in a tonic thermonociceptor. Cell Rep. 2020;30(2):397–408 e4.
81. Nguyen JP, et al. Whole-brain calcium imaging with cellular resolution in freely behaving *Caenorhabditis elegans*. Proc Natl Acad Sci, 2016. 113(8): E1074.
82. Venkatachalam V, et al. Pan-neuronal imaging in roaming *Caenorhabditis elegans*. Proc Natl Acad Sci USA. 2016;113(8):E1082–8.
83. Kotera I, et al. Pan-neuronal screening in *Caenorhabditis elegans* reveals asymmetric dynamics of AWC neurons is critical for thermal avoidance behavior. elife. 2016;5:e19021.
84. Prevedel R, et al. Simultaneous whole-animal 3D imaging of neuronal activity using light-field microscopy. Nat Methods. 2014;11(7):727–30.
85. Sieburth D, Madison JM, Kaplan JM. PKC-1 regulates secretion of neuropeptides. Nat Neurosci. 2007;10(1):49–57.
86. Hawk JD, et al. Integration of plasticity mechanisms within a single sensory neuron of *C. elegans* actuates a memory. Neuron. 2018;97(2):356–367.e4.
87. Nagel G, et al. Light activation of channelrhodopsin-2 in excitable cells of *Caenorhabditis elegans* triggers rapid behavioral responses. Curr Biol. 2005;15(24):2279–84.
88. Schild LC, Glauser DA. Dual color neural activation and behavior control with chrimson and CoChR in *Caenorhabditis elegans*. Genetics. 2015;200(4):1029–34.
89. Edwards SL, et al. A novel molecular solution for ultraviolet light detection in *Caenorhabditis elegans*. PLoS Biol. 2008;6(8):e198.
90. Liu J, et al. *C. elegans* phototransduction requires a G protein-dependent cGMP pathway and a taste receptor homolog. Nat Neurosci. 2010;13(6):715–22.
91. Harbinder S, et al. Genetically targeted cell disruption in *Caenorhabditis elegans*. Proc Natl Acad Sci. 1997;94(24):13128.
92. Chelur DS, Chalfie M. Targeted cell killing by reconstituted caspases. Proc Natl Acad Sci USA. 2007;104(7):2283–8.
93. Xu S, Chisholm AD. Highly efficient optogenetic cell ablation in *C. elegans* using membrane-targeted miniSOG. Sci Rep. 2016;6(1):21271.
94. Macosko EZ, et al. A hub-and-spoke circuit drives pheromone attraction and social behaviour in *C. elegans*. Nature. 2009;458(7242):1171–5.
95. Pokala N, et al. Inducible and titratable silencing of *Caenorhabditis elegans* neurons in vivo with histamine-gated chloride channels. Proc Natl Acad Sci. 2014;111(7):2770.
96. Fang-Yen C, Alkema MJ, Samuel ADT. Illuminating neural circuits and behaviour in *Caenorhabditis elegans* with optogenetics. Philos Trans R Soc Lond Ser B Biol Sci. 2015;370(1677):–20140212.

97. Bergs A, et al. Rhodopsin optogenetic toolbox v2.0 for light-sensitive excitation and inhibition in *Caenorhabditis elegans*. PLoS One. 2018;13(2):–e0191802.
98. Sarma GP, et al. OpenWorm: overview and recent advances in integrative biological simulation of *Caenorhabditis elegans*. Philos Trans R Soc B Biol Sci. 2018;373(1758):20170382.
99. de la Cova C, et al. A real-time biosensor for ERK activity reveals signaling dynamics during *C. elegans* cell fate specification. Dev Cell. 2017;42(5):542–553.e4.
100. Porto D, et al. Conformational changes in twitchin kinase in vivo revealed by FRET imaging of freely moving *C. elegans*. eLife. 2021;10:e66862.
101. Tomida T, et al. The temporal pattern of stimulation determines the extent and duration of MAPK activation in a *Caenorhabditis elegans* sensory neuron. Sci Signal. 2012;5(246):ra76.
102. Motta-Mena LB, et al. An optogenetic gene expression system with rapid activation and deactivation kinetics. Nat Chem Biol. 2014;10(3):196–202.
103. Baaske J, et al. Dual-controlled optogenetic system for the rapid down-regulation of protein levels in mammalian cells. Sci Rep. 2018;8(1):15024.
104. Davis L, et al. Precise optical control of gene expression in *C. elegans* using improved genetic code expansion and Cre recombinase. elife. 2021;10

Regionalization of the Early Nervous System

3

Boris Egger

What You Will Learn in This Chapter

In this chapter, we focus on observations that were made in the field of comparative nervous system development by studies carried out in *Drosophila* and in vertebrate model systems. We will first look at homologous genes that are expressed along the dorsoventral body axis during early neural induction and patterning. Secondly, we discuss findings about the expression and functions of genes involved in anteroposterior patterning of the central nervous system. Despite the significant morphological and developmental differences between insects and vertebrates, there is astonishing conservation between the genetic and molecular mechanisms that control the regionalization of the nervous system. Hence, the data imply that comparable mechanisms operate during embryonic brain development in protostome and deuterostome lineages.

3.1   Introduction

The brain is one of the most complex structures that have ever evolved. Therefore, the molecular mechanisms that control brain development, including the generation of an enormous number of cells in the right position at the right time, are of profound interest. The appearance of brains in different species can vary in many aspects. At a

B. Egger (✉)
Department of Biology, University of Fribourg, Fribourg, Switzerland
e-mail: boris.egger@unifr.ch

first glance, the overall complexity and morphology of the vertebrate central nervous system (CNS) is largely different from the small chain of ganglia that characterize the insect CNS. Until fairly recently, the brains of insects and vertebrates have been considered to be unrelated in their evolutionary origin. There are several reasons for this assumption. First, the developing CNS forms in opposite sides of the embryo in insects as compared to vertebrates. Indeed, bilaterian animals have been subdivided into two major groups based on this difference. The gastroneuralia including protostomes such as insects develop a ventral ladder-like ganglionic chain, called the ventral nerve cord (VNC). In contrast, a dorsal nerve cord characterizes the notoneuralia, which include deuterostomes such as vertebrates. Differences are also seen in the inductive processes that give rise to vertebrate and invertebrate brains during embryogenesis. The vertebrate CNS derives primarily from the dorsal neural plate that invaginates and forms a neural tube. Subsequently, neurogenesis occurs in the neuroepithelium of the closed neural tube where neural stem cells begin to divide. In contrast, in invertebrates, the CNS derives primarily from a single-layered ventral neuroectoderm, from which single neural progenitor cells segregate and start to proliferate. The evolutionary implications of these differences were challenged, however, by the finding of orthologous gene expression and homologous molecular pathways in brain development in vertebrates and invertebrates. Thus, similar molecular mechanisms underlying CNS development were found to operate at the ventral body side in invertebrates and at the dorsal body side in vertebrates. Moreover, in insects and vertebrates, comparable molecular genetic pathways provide positional information for patterning the developing brain along the dorsoventral and anteroposterior body axes.

In 1822, the French naturalist Etienne Geoffrey-Saint-Hilaire dissected a crayfish, placed it upside down, and noted that in this orientation, the organization of the main body system of the crayfish, the nerve cord, muscle, gut, and heart-resembled that of a chordate [1]. From these comparative anatomical studies, he speculated that the dorsoventral axes were inverted in the body plans of vertebrates and arthropods. Over 150 years later, genetic evidence for a conserved developmental program that operates in dorsoventral axis formation in both vertebrates and arthropods was obtained by studying the functions of two sets of homologous genes in *Drosophila* and *Xenopus*. These studies show that the gene homologs *dpp* and *Bmp4* promote dorsal development in *Drosophila* and ventral development in *Xenopus*, whereas the gene homologs *sog* and *Chordin* promote ventral development in the fly *Drosophila* and dorsal development in the frog *Xenopus* [2, 3]. The developmental roles of these molecules have thus been conserved during the more than 500 million years of evolution that separate arthropods and chordates. These results provide strong evidence that in dorsoventral specification the molecular interactions that occur on the ventral side of insects are comparable with those that occur on the dorsal side of vertebrates. These findings on body axis specification open up the possibility that evolutionarily conserved molecular networks might also be responsible for neuraxis specification in insects and vertebrates.

3.1.1 A Morphological Perspective

In *Drosophila*, the embryonic brain is composed of an anterior supraesophageal and an adjacent subesophageal ganglion, and both ganglia derive initially from two separated neurogenic regions. The procephalic neuroectoderm generates the supraesophageal ganglion, which consists of three neuromeric structures, the protocerebrum, the deutocerebrum, and the tritocerebrum. The subesophageal ganglion is derived from the anterior-most ventral neuroectoderm and is also composed of three neuromeres, the mandibular neuromere, the maxillary neuromere, and the labial neuromere (Fig. 3.1a). Posterior to the brain is the ventral nerve cord, which is also derived from the ventral neuroectoderm [4, 7].

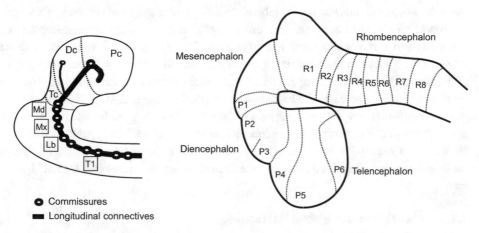

Fig. 3.1 Schematic representation of an embryonic *Drosophila* brain compared to an embryonic vertebrate brain. (**a**) The embryonic *Drosophila* brain is composed of a supraesophageal ganglion and a subesophageal ganlion. The supraesophageal ganglion comprises the protocerebrum (Pc), the deutocerebrum (Dc) and the tritocerebrum (Tc). The subesophageal ganglion comprises the mandibular (Md), maxillary (Mx) and labial (Lb) neuromeres. Posterior to the brain, three thoracic (T1–3) and nine abdominal neuromeres make up the ventral nerve cord. The hemisegment of nearly all neuromeres is interconnected by one or two commissures; the individual neuromeres are connected along the anteroposterior axes by longitudinal connectives. This results in typical ladder-like axon tracts. (**b**) The embryonic vertebrate brain is characterized by a neuromeric organization. It is proposed that the forebrain or prosencephalon is composed of six neuromeric regions called prosomeres (P1–P6). Prosomeres P1–P3 develop into the caudal part of the diencephalon (thalamus) and prosomeres P4–P6 develop into the rostral diencephalon (hypothalamus) and the telencephalon. Posterior to the diencephalon is the midbrain or mesencephalon and the hindbrain or rhombencephalon. The hindbrain becomes subdivided into an anterior metencephalon and a more posterior myelencephalon. Neuromeric subdivisons, called rhombomeres divide the rhombencephalon into smaller compartments (R1–R8). The area posterior to R8 belongs to the embryonic spinal cord [4–6]

About 70–80 neural precursor cells, called neuroblasts, proliferate initially to set up the embryonic brain [8]. Subsequentially, during embryonic development, a simple scaffold of commissural and descending pathways is established by a small set of pioneering axons. This initial set of axons is used for guidance and fasciculation by later outgrowing axons, thus, giving rise to an orthogonal set of commissures and connectives in the embryonic brain. At the level of the protocerebrum and the tritocerebrum, the two bilateral brain hemispheres are interconnected by a prominent preoral commissure and a tritocerebral commissure, respectively (Fig. 3.1a) [4, 7].

In vertebrates, the CNS arises from the dorsal epiblast of the vertebrate gastrula. Inductive interactions between germ layers and the extraembryonic visceral endoderm during gastrulation cause an early specification of the anterior developing neural plate [9]. The neural plate initially becomes morphologically distinct as a thickened layer of epithelial cells and subsequently the neural plate invaginates to form the dorsal neural tube characteristic of the chordate phylum. During these processes, a subdivision takes place, in which the anterior neural tube is dividing into a series of vesicles. This initial anteroposterior patterning results in a prosencephalon or forebrain, a mesencephalon or midbrain, and a rhombencephalon or hindbrain (Fig. 3.1b). The developing hindbrain reveals a clear metameric organization and its regional diversity is achieved by a process of segmentation that bears a superficial resemblance to segmentation in *Drosophila*. The developing rhombomeres of the hindbrain are formed by internal subdivision and reveal pair-wise organization with compartment-like properties [10]. Although the degree of regionalization of the vertebrate prosencephalon is still debated, comparative studies on the restricted expression domains of regulatory genes indicate that this region, like the hindbrain, may be subdivided into neuromeric structures called prosomeres [5, 11].

3.1.2   Neural Dorsoventral Patterning

What are the molecular players, which act during neural induction and define the dorsoventral axis of the developing CNS? In insects as well as in vertebrates, the prospective ectoderm is subdivided into a neurogenic and a non-neurogenic portion by the antagonistic activity of the secreted molecules: Decapentaplegic (Dpp) and Short gastrulation (Sog) in *Drosophila* and their corresponding homologs, Bone morphogenetic protein 4 (BMP4) and Chordin, in vertebrates [12–14]. Additionally, data suggest that beyond the simple formation of the neuroectoderm, significant elements of dorsoventral neuroectoderm patterning have also been evolutionarily conserved. Thus, in flies, three homeobox-containing genes, *ventral nervous system defective* (*vnd*), *intermediate neuroblast defective* (*ind*), and *muscle segment homeobox* (*msh*), are expressed in ventral to dorsal columns in the neuroectoderm. Strikingly, genes that display sequence similarities in vertebrates, belonging to the *Nkx*, *Gsh,* and *Msx* families, are expressed in the neuroepithelium of vertebrates in a comparable manner [15].

3.1.2.1 Morphogenetic Gradients: Sog/Chordin and Dpp/BMP4

In *Drosophila*, most data about the molecular processes of neural tissue formation are obtained from the ventral neuroectoderm. In the early embryo, the germ layers are patterned by the interaction of zygotically transcribed gene products. To date, three major signaling pathways have been shown to play a major role in dorsoventral patterning of neural tissue in *Drosophila*: Dorsal (Dl) signaling is required for ventral mesoderm and neuroectoderm formation, Dpp signaling defines the dorsal border of the neurogenic region and Epidermal growth factor receptor (Egfr) signaling is crucial for the ventral and intermediate neuroectoderm specification [15].

One of the first genes involved in dorsoventral patterning is *dorsal*, which is a member of the Rel/NF kappa B family of transcription factors. Dorsal (Dl) protein is initially distributed throughout the cytoplasm of developing oocytes but is transported into nuclei shortly after fertilization. The Dl protein is distributed in a broad concentration gradient along the dorsoventral axis of the early embryo. This gradient initiates the differentiation of three embryonic tissues, the mesoderm, neuroectoderm, and dorsal ectoderm, by regulating a number of zygotically active target genes in a concentration-dependent manner. High nuclear concentration of Dl at the ventral side of the embryo induces the mesodermal genes *twist* and *snail,* which in turn repress neuroectoderm formation. Lower levels of Dl give rise to neuroectoderm and are required to activate neural gene expression. One of the direct target genes in the neurogenic domain seems to be *sog*. The Dl gradient also functions as a context-dependent repressor that restricts the expression of genes like *dpp* to dorsal regions [16].

The *dpp* gene encodes a member of the transforming growth factor-beta (TGF-beta) superfamily and is observed early in development in the dorsal 40% of the embryonic nuclei. *dpp* expression defines the dorsal border of the presumptive neuroectoderm (Fig. 3.2a) and also has an essential role in establishing the dorsal embryonic tissue including dorsal ectoderm and the extra-embryonic tissue, called the amnioserosa. The *dpp* loss-of-function mutant phenotype shows a marked expansion of the neurogenic ectoderm at the expense of dorsal structures such as the amnioserosa [17]. In contrast, when the *dpp* gene is misexpressed ventrally it can induce dorsal structures and inhibit neurogenic tissue formation [17, 18]. Hence, Dpp acts as an inhibitor of neural tissue formation in the fly.

sog is expressed in broad lateral stripes and is activated by a distinct concentration level within the Dl gradient. Sog is a secreted protein and the initial pattern of *sog* expression appears to coincide with the limits of the presumptive neuroectoderm at the ventral side (Fig. 3.2a). It seems that a morphogenic gradient of Sog antagonizes the dorsalization factor Dpp and prevents the neuroectoderm from becoming dorsal epidermis. Loss of *sog* function causes a reduction of the neuroectoderm with the concomitant expansion of the dorsal epidermis. Interestingly, double mutants of *dpp* and *sog* are indistinguishable from the single *dpp* mutants in their early phenotype. This suggests that *sog* functions through *dpp* [2, 18–20].

Similar to the situation in *Drosophila*, the *dpp* homolog *Bmp4* has a crucial role in dorsoventral patterning in vertebrates (Fig. 3.2b). In *Xenopus*, BMP4 has a strong

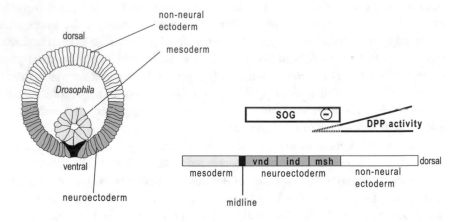

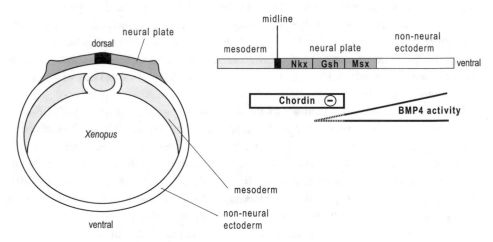

Fig. 3.2 Axis inversion and conserved dorsoventral patterning in *Drosophila* and vertebrates. (**a**) Left: Cross-section through a *Drosophila* embryo at gastrulation stages. The mesoderm invaginates into the inside of the embryo and the neuroectoderm forms at the ventrolateral side. The dorsal side gives rise to non-neural ectoderm and extra-embryonic tissue (amnioserosa). Right: In blastoderm stages, Sog is expressed in ventrolateral cells comprising the neuroectoderm, and Dpp is expressed in dorsal cells comprising the non-neuronal ectoderm. Sog antagonizes Dpp signaling in the neuroectoderm. The genes *vnd*, *ind*, and *msh* pattern the neuroectoderm in three columnar domains. (**b**) Left: Cross-section of a *Xenopus* embryo at the early neurula stage. Vertebrates such as *Xenopus*, have the neural plate positioned at the dorsal body side. Right: BMP4 activity is antagonized by Chordin in the dorsal ectoderm, which leads to the formation of neural tissue. Homeobox genes of the *Nkx*, *Gsh*, and *Msx* families are expressed in a manner similar to *Drosophila* columnar genes. Thus, in both species, dorsoventral ectoderm specification is exerted by homologous DPP/BMP4 and SOG/Chordin action, but at inverse positions. Furthermore, the columnar genes *vnd/Nkx*, *ind/Gsh*, and *msh/Msx* show the inverse spatial order along the dorsoventral axis in vertebrates as compared to *Drosophila* [14, 15]

ventralizing effect and has also been shown to possess anti-neurogenic activity; BMP4 can suppress neural induction and promote the formation of epidermis. When endogenous *Bmp4* signaling is blocked by using either a dominant negative BMP receptor, an antisense *Bmp4* RNA, or a dominant negative form of a BMP4 ligand, neural differentiation is observed even in the absence of organizer-derived neural inducers [21–24]. These data together with the data obtained for *dpp;sog* double mutants in *Drosophila* gave rise to the "neural default" model for ectodermal differentiation [25]. However, data obtained in vertebrates have challenged this model and imply further positive inducers of the neural state [26]. It would be interesting to see if similar additional factors have instructive roles in neural induction in insects.

There is evidence from many transplantation studies that inhibitory signals specify the neural plate during gastrulation. In amphibians and other vertebrates, these instructive signals initially derive from the blastopore lip, also referred to as Spemann's organizer. A number of BMP antagonists are expressed in organizer tissue [14]. Three secreted factors have been identified as neural inducers [27]. Noggin, Follistatin, and Chordin (Fig. 3.2b) can induce neural tissue, which subsequently expresses anterior neural markers such as *Otx2* (see below). Data from *Chordin* studies in *Xenopus* (morpholino injections) demonstrate that downregulation of *Chordin* expression leads to a reduction of neural markers and to expansion of ventral domains [28]. *Chordin* reveals sequence similarity to the *Drosophila sog* gene and in view of their obvious functional similarities, these genes are considered to be homologs. Most likely both of these molecules directly bind to DPP/BMP4 proteins and prevent the binding of these proteins to signaling receptors [29]. Hence, the main function of Chordin and SOG is to inactivate Dpp/Bmp4 signals in the extracellular space and to prevent tissue from becoming epidermal ectoderm.

3.1.2.2 Columnar Genes: *vnd/Nkx*, *ind/Gsh*, and *msh/Msx*

In contrast to information on the specification of the neuroectoderm, less is known about the dorsoventral subdivision of neuroectodermal tissue. Work in *Drosophila* indicates that the newly formed ventral neuroectoderm consists of three dorsoventral domains or columns. Early neural precursors, which are located within one of these three domains, are characterized by the expression of one of the homeobox transcription factors *vnd*, *ind*, and *msh*. *vnd* is expressed in the most ventral domain, *ind* is expressed in an intermediate domain and *msh* defines the most dorsal column (Fig. 3.2a). There is evidence that the *Egfr* pathway helps to control the borders of the intermediate column, since in *Egfr* mutants dorsal genes are expressed in the intermediate neuroectoderm and intermediate neural precursors fail to form [30, 31]. In addition, there seems to be a genetic hierarchy of transcriptional repression among columnar genes in that the more ventral genes repress the more dorsal genes in the domain where they are normally expressed (i.e., *vnd* represses *ind* in the ventral column). In columnar gene mutants, only a few neuroblasts delaminate in the corresponding mutant domain and therefore the activity of the dorsoventral columnar genes appears to be important for the formation and specification of neural progenitor cells [15, 31].

The vertebrate homologs of the *Drosophila* columnar genes show similar patterns of expression in the developing neural plate (Fig. 3.2b). Like *vnd* in the fly, vertebrate *Nkx* family members are expressed in the region of the neural plate nearest to the midline (floorplate) and later specify the most ventral neural cells in the neural tube. Expression of the vertebrate *Msx* family members is found in a lateral position in the neural plate; this is consistent with *msh* expression in the *Drosophila* neuroectoderm. Somewhat in contrast, vertebrate *Gsh* genes, which share sequence identity with *ind*, reveal no expression during neural plate patterning, but these genes are activated later at an intermediate position of the neural tube. Thus, despite some differences in temporal expression patterns, vertebrate and *Drosophila* columnar genes show striking similarities in their spatial order of expression in the neuroectoderm [15].

In vertebrates, it is not clear which factors initiate the dorsoventral subdivision and patterning of the neural plate and might, thus, have a functional role corresponding to the Dorsal and Egfr pathways in *Drosophila* neuroectoderm patterning. One good candidate is *Sonic hedgehog* (*Shh*), which is expressed in the notochord underlying the prospective neuroectoderm and has an essential role in specifying neuronal precursor cells at later stages when the neural tube has already formed [32]. However, it is not clear if *Shh* has an early instructive function in neuroectodermal patterning. Other candidates are BMPs and BMP antagonizers, such as Chordin and Noggin, which might not only function during early dorsoventral patterning of the embryo but could also be involved in patterning of the neural plate at the time of neurulation [14].

The initial determination of neural tissue in insects and vertebrates seems to include highly conserved molecular mechanisms like the antagonistic function of SOG/Chordin versus DPP/BMP4. However, currently only very little is known about genes and genetic networks that link neural induction to primary neurogenesis [33]. An emerging principle of developmental biology is that extracellular gradients initiate the formation of distinct cell types through the differential regulation of target genes, which implement morphogenesis and cell determination. For the Dl gradient in *Drosophila,* genomic studies have identified several target genes by bioinformatic methods and genome-wide oligonucleotide arrays [34, 35]. A similar strategy can be used in vertebrates. Full genomic sequence information for several vertebrate model systems is available and computational methods can be used to identify regulatory regions where transcription factors bind to activate downstream genes [36]. If this is successful, comparative genomics may give us deeper insight into the degree to evolutionary conservation of the gene networks involved in dorsoventral patterning of the CNS in insects and vertebrates.

3.1.3   Anteroposterior Organization of the Developing Brain

What are the molecular mechanisms involved in anteroposterior patterning of the brain? Evidence indicates that a set of evolutionarily conserved homeobox transcription factors is expressed in comparable patterns in the embryonic brains of a wide variety of animal species. Remarkably, genetic experiments carried out in *Drosophila* and in several verte-brate model systems indicate that the functional roles of these transcription factors in brain development also appear to be conserved. In the following, the results of these experiments are summarized with an emphasis on the evolutionary conservation of developmental mechanisms. We focus first on the cephalic gap genes, *orthodenticle* (*otd/Otx*) and *empty spiracle* (*ems/Emx*), which are involved in similar aspects of anterior brain development in insects and vertebrates. Then, we describe the evolutionarily conserved expression and function of the homeotic selector or *Hox* genes, which operate in the development of the posterior brain. Finally, we consider genetic data, which indicates that the developmental genetic control of the boundary region located between the anterior and posterior brain is similar in insects and vertebrates and is likely to be evolutionarily conserved as well.

3.1.3.1 Cephalic Gap Genes Specify the Anterior Brain Territories
otd/Otx Genes
In early blastoderm stages, *orthodenticle* (*otd*) is expressed in a broad circumferential stripe in the anterior region of the embryo. This early expression is required for head development and segmental patterning in *Drosophila* [37–39]. Beside this early function, the *otd* gene is responsible for the specification of the protocerebrum and anterior deutocerebrum in the embryonic *Drosophila* brain. During neurogenesis, *otd* expression is observed throughout most of the protocerebral anlage as well as in the anterior part of the deutocerebral anlage of the brain (Fig. 3.3a) [40, 44]. Additionally, *otd* is expressed along the midline of the developing ventral nerve cord [39, 45].

In the developing brain, mutation of *otd* results in the deletion of the protocerebral anlage due to defective neuroblast formation [40, 44]. In the developing ventral nerve cord, mutational inactivation of *otd* causes CNS differentiation defects in specific midline neurons and glia and this results in deranged or missing commissures [39, 46]. Genetic rescue experiments demonstrate that the gap-like brain defects can be restored in *otd* mutant embryos through ubiquitous overexpression of an *otd* transgene prior to the onset of neurogenesis. Early blastoderm expression is not required for embryonic brain development [47]. The proneural gene *l'sc* is probably a downstream gene of *otd* since in *otd* mutants most of the *l'sc* expression in the protocerebrum and in part in the deutocerebrum is missing. Probably, neuroblasts in the *otd* mutant domain fail to form and, thus, their neuronal progeny in this region is not generated [44].

In mammals, such as the mouse two *otd/Otx* genes have been identified to date. The homeodomains, encoded by *Otx1* and *Otx2*, are very similar and differ only at three and two amino-acid residues from *Drosophila otd* [41]. Detailed molecular and developmental studies indicate that *Otx2* is probably most similar to the fly *otd* gene [48]. In the mouse, *Otx2* is expressed from the earliest embryonic stages onward in the epiblast and in the extraembryonic visceral endoderm. At the end of gastrulation, *Otx2* is also expressed in the most anterior neural plate which gives rise to the telencephalon, diencephalon, and mesencephalon (Fig. 3.3b). Patterning of the anterior neuroectoderm in vertebrates is thought to involve two sequential steps [49]. First, the induction of the anterior neural plate

A. *Drosophila*

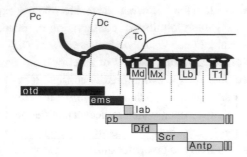

B. Vertebrate

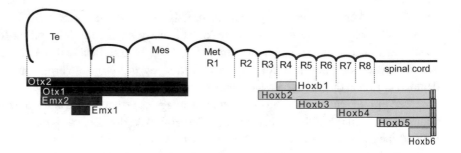

C. Hox Clusters

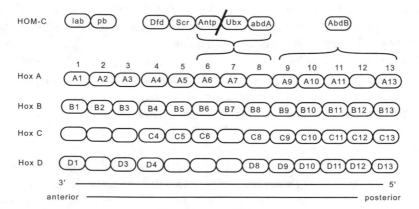

Fig. 3.3 Conserved anteroposterior patterning in the embryonic brain. and the *HOM/Hox* clusters. (**a**) Schematic representation of gap and homeotic gene expression in the embryonic brain of *Drosophila*. Gene expression domains are delimited by labeled bars. The axonal scaffold in hemisegments is outlined (black). The two hemisegment of nearly all neuromeres are interconnected by commissures and the individual neuromeres are connected along the anteroposterior axes by longitudinal connectives. Neuromere boundaries are indicated by dotted lines. Brain neuromeres

and second the regional specification of forebrain and midbrain areas, and *Otx2* appears to be involved in both of these events. Complete *Otx2* null mutant embryos die early in embryogenesis, lack the anterior neuroectoderm fated to become forebrain, midbrain, and anterior hindbrain, and show strong abnormalities in their body plan. As a result, this can lead to a dramatic headless phenotype [50–52].

The onset of expression of *Otx1* occurs later than that of *Otx2*, and *Otx1* expression covers a region that broadly overlaps with that of *Otx2*. In contrast to the early expression of *Otx2*, which disappears from the dorsal telencephalon, *Otx1* expression is maintained uniformly across the ventricular zone of the cortical anlagen. During neurogenesis, *Otx1* expression is found in cells that generate neurons of layers 5 and 6 in the cortex [53, 54]. Absence of *Otx1* function leads to a significant reduction of cell number in the cerebral cortex and other brain regions. Probably defective proliferation of neuronal progenitors is responsible for the overall reduction of *Otx1* mutant brains. In addition, these mutant mice suffer from spontaneous epileptic attacks [55].

Otx1 and *Otx2* can partially rescue each other's mutant phenotype as demonstrated by genetic replacement studies. In *Otx1* mutant mice, the human *OTX2* gene can fully restore corticogenesis abnormalities and epilepsy phenotype. On the other hand, when *Otx2* is replaced with human *OTX1* (*hOTX1*) the anterior neural plate induction is restored but these mice still show a headless phenotype, suggesting that the early patterning of the neuroectoderm is recovered but that maintenance of forebrain-midbrain identities is still lost [56]. In this context, it is noteworthy that although the mRNA of *hOTX1* is found in both visceral endoderm and epiblast, the hOTX1 protein is detected only in the visceral endoderm. In contrast, *Otx2* mRNA is translated in all of the cells, in which it is expressed. This suggests a differential translational regulation of *Otx1* versus *Otx2* mRNA. Indeed, the untranslated region of *Otx2* possesses sequence specificity which mediates translation in epiblast, axial mesendoderm, and anterior neuroectoderm but which is not required for translation in the visceral endoderm [57].

Fig. 3.3 (continued) from anterior to posterior are protocerebrum (Pc), deutocerebrum (Dc), tritocerebrum (Tc), mandibular neuromere (Md), maxillary neuromere (Mx), labial neuromere (Lb) and first thoracic neuromere (T1). (**b**) Schematic representation of gap and homeotic gene expression in the embryonic brain of a vertebrate (mouse). Labeled bars delimit gene expression domains. Dotted lines delimit telencephalon (Te), diencephalon (Di), mesencephalon (Mes), metencephalon (Met), and rhombomeres (R1–R8). (**c**) *Hox* gene clusters. *HOM-C* in *Drosophila* contains eight genes and is separated into two complexes (slashed line). In vertebrates 39 *Hox* genes were identified which are arranged in four paralogous clusters (*Hox A–D*). In *Drosophila* and in vertebrates, the 3′ (proximal) to 5′ (distal) gene order along the chromosome corresponds to an anteroposterior spatial co-linearity of expression in the CNS. Exceptions are *pb/Hoxb2* genes, which are expressed anteriorly to *lab/Hoxb1* genes [40–43]

ems/Emx Genes

A second cephalic gap gene, *empty spiracles* (*ems*), is responsible for the specification of the posterior deutocerebrum and anterior tritocerebrum in the embryonic *Drosophila* brain. Similar to the *otd* gene, *ems* is expressed at the blastoderm stage in a single circumferential stripe at the anterior end of the embryo. Later in embryogenesis, *ems* is expressed in the developing procephalic region that gives rise to the deutocerebral and tritocerebral anlagen (Fig. 3.3a). In addition to its expression in the head region, *ems* also shows a later, metameric expression pattern in ectodermal and neural cell patches in all trunk segments [58–60]. Mutational inactivation of *ems* leads to the deletion of the deutocerebral and tritocerebral anlagen of the embryonic brain [40]. This is probably due to defective specification of the neuroectoderm in these anlagen resulting in the absence of neuroblasts, and this, in turn results in the lack of most of the deutocerebral and the tritocerebral neurons. As in the case of *otd* null mutants, the absence of neuroblasts in the *ems* mutant domains correlates with the loss of proneural gene expression [44]. Interestingly, rescue of the *ems* null mutant phenotype can be achieved by ubiquitously overexpressing *ems* at an embryonic stage, when in the wild-type most of the *ems* expressing neuroblasts have already delaminated. This overexpression results in a restoration of brain morphology, indicating that *ems* expression at this stage is sufficient for proper brain development [60].

The mouse *Emx* gene family consists of two members, *Emx1* and *Emx2* [61, 62]. Within their homeodomains, murine EMX1 and EMX2 proteins differ from *Drosophila* EMS protein in 11 aminoacids. The *Emx* genes are predominately expressed in regions of the developing anterior brain [41, 63, 64]. *Emx1* is restricted to the dorsal telencepahlon with its posterior expression boundary slightly anterior to the boundary between the presumptive telencephalon and diencephalon (Fig. 3.3b). At later developmental stages, *Emx1* is expressed in most cortical neurons, regardless of whether they are proliferating, migrating, or differentiating. *Emx1* mutant mice survive, show no obvious behavioral defects, and reveal subtle defects restricted to the forebrain. For example, the cortical plate and the white matter of mutants are thinner than normal, and the subplate is hardly visible [65].

Emx2 is expressed both in the dorsal and ventral telencephalon and in the hypothalamus and is one of the earliest markers for the developing cerebral cortex. Its expression domain is delimited by a posterior boundary, which is located within the roof of the presumptive diencepahlon (Fig. 3.3b). During corticogenesis, *Emx2* transcripts are found in the neuroepithelium but are absent from post-mitotic neurons of the cortical plate. *Emx2* expression covers the ventricular zone during neurogenesis, and it forms a gradient with highest expression levels in regions with a low proliferation rate [64]. Additionally, *Emx2* is expressed in Cajal–Retzius cells, a transient cell population of the first-born neurons in the neocortex that forms the most-superficial cell layer underlying the pial membrane. In accordance with this expression domain, the Cajal–Retzius cells are lacking in *Emx2* mutant embryos. As a consequence, the settling of radial glia is disturbed and neurons display abnormal migration patterns; late-born neurons largely fail to pass early-born neuron layers and thus the characteristic inside-out layering is disrupted [66]. *Emx2* expression in the proliferative layer of the developing cortex and in the Cajal–Retzius cells suggests a potential role in regulating proliferation, migration, and differentiation. Indeed, *Emx2* null embryonic brains have major abnormalities in the architecture of various brain regions, including the cerebral cortex. Furthermore, studies *in vivo* and with embryonic and adult stem cells in vitro demonstrate a functional role of *Emx2* in cell fate decisions and in the transition from symmetric to asymmetric cell division [67–69].

Conserved Molecular Function of Cephalic Gap Genes Cross-Phylum Rescue Experiments

Investigations of expression patterns and mutant phenotypes in the embryonic brains of *Drosophila* and mouse reveal developmental mechanisms that are strikingly similar and suggest an evolutionary conservation of *otd/Otx* and *ems/Emx* genes in the embryonic brain. In addition, cross-phylum gene rescue experiments provide remarkable evidence for the evolutionary conservation of functional properties of cephalic gap genes. Ubiquitous overexpression of human *OTX2* in the *otd* null mutant background is able to restore the anterior brain morphology in *Drosophila* embryos. Ubiquitous overexpression of human *OTX1* in the *otd* null mutant background is also able to rescue anterior brain morphology, however with less efficiency. These data suggest that vertebrate *Otx* genes have the capability to replace the *otd* gene in the development of the anterior part of the embryonic fly brain [47]. It is noteworthy that also ubiquitous overexpression of the ascidian *Otx* gene is able to replace the *Drosophila otd* gene in the formation of the anterior embryonic brain [70]. Are reciprocal cross-phylum rescue experiments possible in vertebrates? To determine if *Drosophila otd* can substitute for *Otx* genes in anterior brain development of the mouse, Acampora and colleagues replaced the mouse *Otx1* gene or the mouse *Otx2* gene with the fly *otd* gene (cDNA). In these experiments, the *Drosophila otd* could rescue most *Otx1* mutant phenotypes. The rescued adult mice show similar brain size as well as thickness and cell number in cortical layers as seen in the wild-type, suggesting that normal proliferation activity is restored in the dorsal telencephalic neuroepithelium [56]. In contrast, when the mouse *Otx2* gene was replaced in the same way with *Drosophila otd* (cDNA), the *otd* mRNA could be detected in the anterior visceral endoderm and in the epiblast, but Otd protein was only present in the visceral endoderm. (Similar findings were obtained in experiments in which *Otx2* was replaced with *hOTX1*). In consequence, both the early defects in gastrulation and anterior neural plate formation could be restored; however, later these embryos failed to maintain the anterior-most identities of the brain and became headless. In subsequent experiments, Acampora and colleagues demonstrated that untranslated regions of the *Otx2* locus are required for proper translation in epiblast and neuroectoderm. They generated a mouse wherein only the protein-coding sequences (and introns) of *Otx2* were replaced by *otd* coding sequences, but the endogenous mouse flanking sequences encoding the 5′ and 3′ UTRs were preserved. In these mouse models, *otd* mRNA with flanking mouse UTRs was translated properly in epiblast and neuroectoderm and later maintenance and patterning of the forebrain was restored [71]. This analysis suggests that *Drosophila* Otd and mouse

OTX2 proteins share functional equivalence, however, the regulation of *otd/Otx* genes has been modified during evolution. Indeed, the *Otx2* 3' UTR of all gnathostome vertebrates analyzed reveal a highly conserved element whereas in protochordates, echinoderms and insects no significant sequence similarities were found to this element [57].

Cross-phylum replacement experiments were also performed for the *ems/Emx* genes. When a mouse *Emx2* transgene was overexpressed in an *ems* null mutant background in *Drosophila* a substantial rescue efficiency in embryonic brain development was achieved [60]. Corresponding cross-phylum rescue experiments have not yet been carried out in the mouse, and therefore it is not known if the *Drosophila ems* gene is able to restore the defects observed in mouse *Emx2* mutants. Despite the incomplete information on the cross-phylum rescue potential of the *ems/Emx* genes, the overall result of the cross-phylum rescue experiments performed so far suggests that the function of *otd/Otx* and *ems/Emx* genes in brain development are to a large degree evolutionary conserved between protostomes and deuterostomes.

3.1.3.2 Homeotic Selector (*Hox*) Genes Specify Posterior Hindbrain Neuromeres

The homeotic or *Hox* genes were originally discovered in *Drosophila* through the homeotic transformation that resulted from their mutation [72]. The homeotic genes show a spatial co-linearity in their chromosomal arrangement and their expression patterns, in that more 3' located genes are expressed more anteriorly along the body axis of the embryo, whereas more 5' located genes are expressed more posteriorly (Fig. 3.3). Furthermore, there appears to be a functional hierarchy among *Hox* gene products in that more posteriorly expressed *Hox* genes are functionally dominant over more anteriorly expressed Hox genes; this is termed "posterior prevalence" [73]. In *Drosophila*, the homeotic genes are arranged in one cluster, but map to the separated Antennapedia (ANT-C) and Bithorax (BX-C) complexes, which are collectively referred to as the Homeotic complex (HOM-C). The ANT-C includes the genes *labial* (*lab*), *probpscipedia* (*pb*), *Deformed* (*Dfd*), *Sex combs reduced* (*Scr*), and *Antennapedia* (*Antp*). The BX-C contains the genes *Ultrabithorax* (*Ubx*), *abdominal-A* (*abd-A*), and *Abdominal-B* (*Abd-B*) (Fig. 3.3c) [74].

In the embryonic brain of *Drosophila*, homeotic gene expression is not observed in the most anterior regions where *otd* and *ems* is required for neuromere formation. (With the exception of *pb*, which has an atypical expression pattern.) The homeotic gene with the most anterior expression domain in the embryonic brain is *lab*, which is expressed in the posterior tritocerebrum. *Lab* expression is followed by non-overlapping domains of *Dfd*, *Scr*, and *Antp* expression in the mandibular, maxillary and labial neuromeres, respectively. The BX-C genes are expressed in the more posterior thoracic and abdominal neuromeres. Neuroanatomical analyses have shown that loss-of-function mutations for two Hox genes, *lab* and *Dfd*, result in severe defects of the embryonic brain. In *lab* null mutants, the neural progenitor cells that give rise to the tritocerebrum as well as their postmitotic progenies are present and correctly positioned in the mutant domain. However, these cells do not extend axons or dendrites and are not contacted by axons from other parts of the brain.

Furthermore, they do not express the neuron-specific markers that positionally equivalent neuronal cells express in the wild type. Thus, they seem to remain in an undifferentiated state and fail to adopt a neuronal identity. It results in axonal patterning defects including loss of the tritocerebral commissure and reduced or absent longitudinal pathways; these are due to both cell-autonomous and cell-non-autonomous effects. Glial cell differentiation appears to be unaffected in the mutant domain. Similar phenotypes were observed in *Dfd* mutants in the corresponding mandibular neuromere, but not in loss-of-function mutants for other homeotic genes [42].

In vertebrates, Hox genes show striking similarities in structure, chromosomal organization, and CNS expression domains as compared to the homeotic genes of *Drosophila*. However, unlike in HOM-C, the mouse Hox genes are grouped in four paralogous clusters (termed A-D), which are presumably derived by duplication of an ancestral cluster. Thus, while the *Drosophila* homeotic complex only comprises 8 genes, there are 39 Hox genes in the mouse (Fig. 3.3c) [72]. Analysis of Hox gene expression patterns in mouse embryos has shown that the Hox genes are expressed in partially overlapping domains along the anteroposterior axis of the developing hindbrain and spinal cord [43, 75]. Like the HOM-C genes in *Drosophila*, the genes most closely related to ANT-C genes have expression domains, which extend to more anterior rhombomere or presumptive spinal cord regions whereas BX-C-related genes are expressed in the more posterior regions of the presumptive spinal cord (Fig. 3.3b).

Two Hox genes related to the *lab* gene, Hoxa1 and Hoxb1 are expressed in rhombomere 4 (R4). The expression domains of both genes reach a sharp anterior boundary in the neuroectoderm coinciding with the presumptive R3/R4 border. Mutational inactivation of Hoxa1 results in segmentation defects that lead to the partial deletion of rhombomeres, suggesting a role in generating and/or maintaining segmental compartments. Hoxb1 loss-of-function produces an alteration in R4 identity, which leads to aberrant migration of neurons within this domain and to a partial transformation of R4 into an R2 identity [76]. In Hoxa1−/−; Hoxb1−/− double mutants, a region corresponding to R4 is formed, but R4-specific markers fail to be activated, indicating the presence of a territory between R3 and R5 with an unknown identity. Furthermore, these double mutants display a reduced number of a specified neuron type whose axons now appear to exit randomly from the neural tube without fasciculations toward distinct exit points [77, 78]. These results are reminiscent of the *lab* mutant phenotypes observed in the embryonic brain of *Drosophila* and suggest remarkably similar mechanisms of lab/Hox1 gene action in specifying neuromere-specific neuronal identity in insects and vertebrates.

3.1.3.3 Tripartite Organization of the Urbilaterian Brain

In the brain of vertebrates, a midbrain–hindbrain boundary (MHB) or isthmus is located between the presumptive mesencephalon and metencephalon. During embryogenesis, the MHB develops in a region that is intercalated between the anterior Otx2 expression domain and the posterior Hox expression domains. The MHB region has organizer-like features since tissue from this region transplanted to the diencephalon or the rhombencephalon induces the cells surrounding it to develop mesencephalic fates in the diencephalon or cerebellar fates in the rhombencephalon. A number of transcription factors and signaling molecules are expressed in the presumptive MHB region, and these generate a complex genetic network, which is required to establish and to maintain the organizer-like properties of this domain [79–81]. During gastrulation and early neurulation, Otx2 and another homeobox gene, gastrulation brain homeobox 2 (Gbx2), are expressed in mutually exclusive domains that lie anterior and posterior to the presumptive MHB. When the level of Otx gene expression is genetically reduced, the anterior margin of Gbx2 expression shifts to a more anterior position. Conversely, in Gbx−/− mutant brains, Otx2 expression is shifted to a more posterior position. These and other experiments indicate that Otx2 and Gbx2 control the induction and positioning of the MHB [82, 83]. At the interface of the Otx2 and Gbx2 expression domains, genes encoding transcription factors of the Pax2/5/8 families delimit an intermediate expression domain that coincides with the presumptive MHB (Fig. 3.4). Mutational inactivation of either Pax2 or Pax5 alone or a combination of double mutants leads to partial or complete deletion of midbrain and cerebellum structures [79–81].

To investigate whether the insect brain also possesses an MHB-like boundary region intercalated between the anterior *otd* expression domain and the posterior homeotic expression domains, Hirth and colleagues have carried out a comparative analysis of expression and function for the fly orthologs of the genes that pattern the vertebrate MHB region [84]. The *Drosophila* genome contains two genes, *Pox neuro* (*Pox-n*) and *Drosophila Pax2* (*DPax2*), that are together considered to be orthologs of the *Pax2/5/8* group [85] and one *Gbx2* orthologous gene called *unplugged* (*unpg*) [86]. Remarkably, the *Pax2/5/8* orthologs *Pox-n* and *DPax2* are expressed in the embryonic fly brain in a transversal boundary region located posterior to the *otd* expression domain in the deutocerebral neuromere and anterior to the homeotic gene expression domain in the tritocerebral neuromere. Furthermore, the *Gbx2* ortholog *unpg* reveals its anteriormost expression in the posterior deutocerebrum, which coincides with the posterior border of *otd* expression. There are no cells found that co-express *otd* and *unpg* and thus, these domains exclude each other. Comparable to the situation in the vertebrates, in *otd* loss-of-function mutants *unpg* expression is shifted anteriorly. Null mutants for *unpg*, in contrast, reveal posteriorly shifted *otd* expression extending into the posterior deutocerebrum. Thus, in both *Drosophila* and vertebrates the *otd/Otx2* or *unpg/Gbx2* genes appear to negatively regulate each other at the interface of their expression domains in a boundary zone located between anterior and posterior brain regions. In contrast to vertebrates, mutational inactivation of the *Pox-n* or *DPax2* does not result in obvious brain defects in *Drosophila* [84]. Nevertheless, the striking similarities at the level of gene expression and functional interactions of *otd/Otx*, *unpg/Gbx2*, *Pax2/5/8*, and *Hox* genes represent a further example of evolutionary conservation of brain patterning mechanisms. Moreover, these results

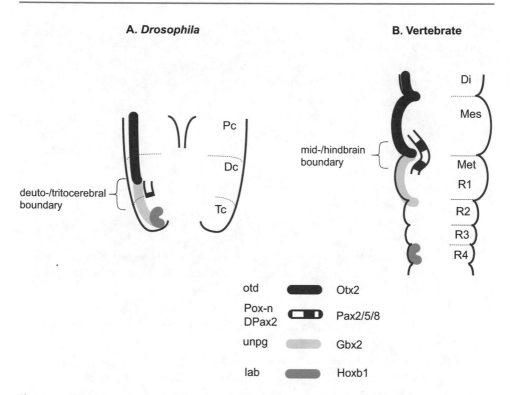

A. *Drosophila* **B. Vertebrate**

Fig. 3.4 Tripartite organization of the *Drosophila* and vertebrate brain. Simplified summary scheme of the expression of the genes *otd/Otx2*, *unpg/Gbx2*, *Pax2/5/8*, and *lab/Hoxb1* in the developing CNS of *Drosophila* (**a**) and a mouse (**b**). (**a**) Schematic representation of an early embryonic *Drosophila* brain, dorsal view, anterior to the top. The protocerebral (Pc), deutocerebral (Dc) and tritocerebrbral (Tc) neuromeres are shown. In *Drosophila,* the *Pax2/5/8* genes, *Pox-n* and *DPax2*, are both expressed in a transversal boundary region at the interface of the *otd* expression domain and the *unpg* expression domain and anterior to the *lab* expression domain. (**b**) Schematic representation of an early embryonic mouse brain, dorsal view, anterior to the top. The diencephalon (Di), mesencephalon (Mes), metencephalon (Met), and first four rhombomercs (R1–R4) arc shown. Similar to *Drosophila* the *Pax2/5/8* genes are expressed in a transversal boundary region at the interface of the *Otx2* expression domain and the *Gbx2* expression domain and anterior to the *Hoxb1* expression domain. In vertebrates, this region forms the midbrain/hindbrain boundary and has organizer-like features [79, 84]

suggest that a tripartite organization of the brain (anterior brain, boundary region, posterior brain) was already present in a urbilaterian ancestor before the separation of protostome and deuterostome lineages.

Question

Describe the hypothesis of dorsoventral axis inversion? What is the evidence that supports this hypothesis?

Take-Home Message

In this chapter, we focused on developmental and molecular mechanisms, which reveal similarities in nervous system and in brain development between *Drosophila* and vertebrates.

- Conserved dorsoventral patterning mechanisms involve morphogenetic factors such as DPP/BMP4 and Sog/Chordin, and the "columnar" genes *vnd/Nkx*, *ind/Gsh*, *msh/Msx*.
- Evidence shows that in insects and vertebrates, genes of the *otd/Otx* family and *ems/Emx* family are responsible for the development of the anterior brain, whereas the *Hox* genes are essential for posterior brain patterning.
- The *Pax 2/5/8* gene family and *unpg/Gbx2* are expressed in similar manner in an intermediate region between the anterior *Otx* and posterior *Hox* domains. Thus, homologous developmental control genes are involved in anteroposterior patterning of the brain in insects and vertebrates.

One of the most remarkable endeavors in evolutionary biology is the developmental reconstruction of the last common urbilaterian ancestor of protostomes and deuterostomes. The striking similarities in gene expression, gene function and gene interchangeability in the neural and brain development of insects and vertebrates presented in this chapter imply the existence of a widely conserved genetic mechanisms for constructing the nervous system, including the brain. In consequence, it is possible that the divergent brain types found in present animals as diverse as flies and humans might have a common evolutionary origin in a brain-like structure that existed before the protostome-deuterostome split over 570 million years ago. However, similarities in gene expression and in functional hierarchies among genetic pathways may not be enough to make such conclusions [87]. The demonstration that entire complex brain regions in insects and vertebrates are composed of developmentally and genetically similar units would clearly add weight to the notion of a universal blueprint for building the brain. In this sense, it will continue to be a challenge for current and future research to unravel the developmental and molecular mechanisms that give rise to the complex multicellular structures that make up the brains in bilaterian animals.

Acknowledgment I am grateful to Professor Dr. Heinrich Reichert (*1949, †2019) for his advice and discussions toward this chapter.

References

1. Arendt D, Nübler-Jung K. Inversion of dorsoventral axis? Nature. 1994;371(6492):26.
2. Holley SA, Jackson PD, Sasai Y, Lu B, De Robertis EM, Hoffmann FM, et al. A conserved system for dorsal-ventral patterning in insects and vertebrates involving *sog* and *chordin*. Nature. 1995;376(6537):249–53.
3. Schmidt J, Francois V, Bier E, Kimelman D. *Drosophila short gastrulation* induces an ectopic axis in *Xenopus*: evidence for conserved mechanisms of dorsal-ventral patterning. Development. 1995;121(12):4319–28.
4. Therianos S, Leuzinger S, Hirth F, Goodman CS, Reichert H. Embryonic development of the *Drosophila* brain: formation of commissural and descending pathways. Development. 1995;121(11):3849–60.
5. Rubenstein JL, Martinez S, Shimamura K, Puelles L. The embryonic vertebrate forebrain: the prosomeric model. Science. 1994;266(5185):578–80.
6. Lumsden A, Krumlauf R. Patterning the vertebrate neuraxis. Science. 1996;274(5290):1109–15.
7. Nassif C, Noveen A, Hartenstein V. Embryonic development of the *Drosophila* brain. I. Pattern of pioneer tracts. J Comp Neurol. 1998;402(1):10–31.
8. Younossi-Hartenstein A, Nassif C, Green P, Hartenstein V. Early neurogenesis of the *Drosophila* brain. J Comp Neurol. 1996;370(3):313–29.
9. Beddington RS, Robertson EJ. Anterior patterning in mouse. Trends Genet. 1998;14(7):277–84.
10. Lumsden A. The cellular basis of segmentation in the developing hindbrain. Trends Neurosci. 1990;13(8):329–35.
11. Rubenstein JL, Shimamura K, Martinez S, Puelles L. Regionalization of the prosencephalic neural plate. Annu Rev Neurosci. 1998;21:445–77.
12. Bier E. Anti-neural-inhibition: a conserved mechanism for neural induction. Cell. 1997;89(5):681–4.
13. Ferguson EL. Conservation of dorsal-ventral patterning in arthropods and chordates. Curr Opin Genet Dev. 1996;6(4):424–31.
14. Sasai Y, De Robertis EM. Ectodermal patterning in vertebrate embryos. Dev Biol. 1997;182(1):5–20.
15. Cornell RA, Von Ohlen T. vnd/Nkx, ind/Gsh, and msh/Msx: conserved regulators of dorsoventral neural patterning? Curr Opin Neurobiol. 2000;10(1):63–71.
16. Stathopoulos A, Levine M. Dorsal gradient networks in the *Drosophila* embryo. Dev Biol. 2002;246(1):57–67.
17. Wharton KA, Ray RP, Gelbart WM. An activity gradient of *decapentaplegic* is necessary for the specification of dorsal pattern elements in the *Drosophila* embryo. Development. 1993;117(2):807–22.
18. Ferguson EL, Anderson KV. Decapentaplegic acts as a morphogen to organize dorsal-ventral pattern in the *Drosophila* embryo. Cell. 1992;71(3):451–61.
19. Francois V, Solloway M, O'Neill JW, Emery J, Bier E. Dorsal-ventral patterning of the *Drosophila* embryo depends on a putative negative growth factor encoded by the *short gastrulation* gene. Genes Dev. 1994;8(21):2602–16.
20. Biehs B, Francois V, Bier E. The *Drosophila short gastrulation* gene prevents Dpp from autoactivating and suppressing neurogenesis in the neuroectoderm. Genes Dev. 1996;10(22):2922–34.
21. Suzuki A, Shioda N, Ueno N. Bone morphogenetic protein acts as a ventral mesoderm modifier in early *Xenopus* embryos. Develop Growth Differ. 1995;37:581–8.
22. Hawley SH, Wunnenberg-Stapleton K, Hashimoto C, Laurent MN, Watabe T, Blumberg BW, et al. Disruption of BMP signals in embryonic *Xenopus* ectoderm leads to direct neural induction. Genes Dev. 1995;9(23):2923–35.

23. Xu RH, Kim J, Taira M, Zhan S, Sredni D, Kung HF. A dominant negative bone morphogenetic protein 4 receptor causes neuralization in *Xenopus* ectoderm. Biochem Biophys Res Commun. 1995;212(1):212–9.

24. Wilson PA, Hemmati-Brivanlou A. Vertebrate neural induction: inducers, inhibitors, and a new synthesis. Neuron. 1997;18(5):699–710.

25. Hemmati-Brivanlou A, Melton D. Vertebrate neural induction. Annu Rev Neurosci. 1997;20:43–60.

26. Bally-Cuif L, Hammerschmidt M. Induction and patterning of neuronal development, and its connection to cell cycle control. Curr Opin Neurobiol. 2003;13(1):16–25.

27. Thomsen GH. Antagonism within and around the organizer: BMP inhibitors in vertebrate body patterning. Trends Genet. 1997;13(6):209–11.

28. Oelgeschlager M, Kuroda H, Reversade B, De Robertis EM. *Chordin* is required for the Spemann organizer transplantation phenomenon in *Xenopus* embryos. Develomental Cell. 2003;4(2):219–30.

29. Piccolo S, Sasai Y, Lu B, De Robertis EM. Dorsoventral patterning in *Xenopus*: inhibition of ventral signals by direct binding of chordin to BMP-4. Cell. 1996;86(4):589–98.

30. Udolph G, Urban J, Rusing G, Luer K, Technau GM. Differential effects of EGF receptor signalling on neuroblast lineages along the dorsoventral axis of the *Drosophila* CNS. Development. 1998;125(17):3291–9.

31. Von Ohlen T, Doe CQ. Convergence of *dorsal*, *dpp*, and *Egfr* signaling pathways subdivides the *Drosophila* neuroectoderm into three dorsal-ventral columns. Dev Biol. 2000;224(2):362–72.

32. Marti E, Bovolenta P. Sonic hedgehog in CNS development: one signal, multiple outputs. Trends Neurosci. 2002;25(2):89–96.

33. Sasai Y. Regulation of neural determination by evolutionarily conserved signals: anti-BMP factors and what next? Curr Opin Neurobiol. 2001;11(1):22–6.

34. Markstein M, Markstein P, Markstein V, Levine MS. Genome-wide analysis of clustered Dorsal binding sites identifies putative target genes in the *Drosophila* embryo. Proc Natl Acad Sci USA. 2002;99(2):763–8.

35. Stathopoulos A, Van Drenth M, Erives A, Markstein M, Levine M. Whole-genome analysis of dorsal-ventral patterning in the *Drosophila* embryo. Cell. 2002;111(5):687–701.

36. Markstein M, Levine M. Decoding cis-regulatory DNAs in the *Drosophila* genome. Curr Opin Genet Dev. 2002;12(5):601–6.

37. Gao Q, Wang Y, Finkelstein R. *Orthodenticle* regulation during embryonic head development in *Drosophila*. Mech Dev. 1996;56(1–2):3–15.

38. Cohen SM, Jürgens G. Mediation of *Drosophila* head development by gap-like segmentation genes. Nature. 1990;346(6283):482–5.

39. Finkelstein R, Smouse D, Capaci TM, Spradling AC, Perrimon N. The *orthodenticle* gene encodes a novel homeo domain protein involved in the development of the *Drosophila* nervous system and ocellar visual structures. Genes Dev. 1990;4(9):1516–27.

40. Hirth F, Therianos S, Loop T, Gehring WJ, Reichert H, Furukubo-Tokunaga K. Developmental defects in brain segmentation caused by mutations of the homeobox genes *orthodenticle* and *empty spiracles* in *Drosophila*. Neuron. 1995;15(4):769–78.

41. Simeone A, Acampora D, Gulisano M, Stornaiuolo A, Boncinelli E. Nested expression domains of four homeobox genes in developing rostral brain. Nature. 1992;358(6388):687–90.

42. Hirth F, Hartmann B, Reichert H. Homeotic gene action in embryonic brain development of *Drosophila*. Development. 1998;125(9):1579–89.

43. Keynes R, Krumlauf R. Hox genes and regionalization of the nervous system. Annu Rev Neurosci. 1994;17:109–32.

44. Younossi-Hartenstein A, Green P, Liaw GJ, Rudolph K, Lengyel J, Hartenstein V. Control of early neurogenesis of the *Drosophila* brain by the head gap genes *tll*, *otd*, *ems*, and *btd*. Dev Biol. 1997;182(2):270–83.
45. Wieschaus E, Perrimon N, Finkelstein R. *orthodenticle* activity is required for the development of medial structures in the larval and adult epidermis of *Drosophila*. Development. 1992;115(3):801–11.
46. Klämbt C, Jacobs JR, Goodman CS. The midline of the *Drosophila* central nervous system: a model for the genetic analysis of cell fate, cell migration, and growth cone guidance. Cell. 1991;64(4):801–15.
47. Leuzinger S, Hirth F, Gerlich D, Acampora D, Simeone A, Gehring WJ, et al. Equivalence of the fly *orthodenticle* gene and the human *OTX* genes in embryonic brain development of *Drosophila*. Development. 1998;125(9):1703–10.
48. Acampora D, Gulisano M, Broccoli V, Simeone A. *Otx* genes in brain morphogenesis. Prog Neurobiol. 2001;64(1):69–95.
49. Stern CD. Initial patterning of the central nervous system: how many organizers? Nat Rev Neurosci. 2001;2(2):92–8.
50. Ang SL, Jin O, Rhinn M, Daigle N, Stevenson L, Rossant J. A targeted mouse *Otx2* mutation leads to severe defects in gastrulation and formation of axial mesoderm and to deletion of rostral brain. Development. 1996;122(1):243–52.
51. Matsuo I, Kuratani S, Kimura C, Takeda N, Aizawa S. Mouse *Otx2* functions in the formation and patterning of rostral head. Genes Dev. 1995;9(21):2646–58.
52. Acampora D, Mazan S, Lallemand Y, Avantaggiato V, Maury M, Simeone A, et al. Forebrain and midbrain regions are deleted in *Otx2−/−* mutants due to a defective anterior neuroectoderm specification during gastrulation. Development. 1995;121(10):3279–90.
53. Frantz GD, Weimann JM, Levin ME, McConnell SK. *Otx1* and *Otx2* define layers and regions in developing cerebral cortex and cerebellum. J Neurosci. 1994;14(10):5725–40.
54. Simeone A, Acampora D, Mallamaci A, Stornaiuolo A, D'Apice MR, Nigro V, et al. A vertebrate gene related to *orthodenticle* contains a homeodomain of the bicoid class and demarcates anterior neuroectoderm in the gastrulating mouse embryo. EMBO J. 1993;12(7):2735–47.
55. Acampora D, Mazan S, Avantaggiato V, Barone P, Tuorto F, Lallemand Y, et al. Epilepsy and brain abnormalities in mice lacking the *Otx1* gene. Nat Genet. 1996;14(2):218–22.
56. Acampora D, Avantaggiato V, Tuorto F, Briata P, Corte G, Simeone A. Visceral endoderm-restricted translation of *Otx1* mediates recovery of *Otx2* requirements for specification of anterior neural plate and normal gastrulation. Development. 1998;125(24):5091–104.
57. Pilo Boyl P, Signore M, Acampora D, Martinez-Barbera JP, Ilengo C, Annino A, et al. Forebrain and midbrain development requires epiblast-restricted *Otx2* translational control mediated by its 3' UTR. Development. 2001;128(15):2989–3000.
58. Dalton D, Chadwick R, McGinnis W. Expression and embryonic function of *empty spiracles*: a *Drosophila* homeo box gene with two patterning functions on the anterior-posterior axis of the embryo. Genes Dev. 1989;3(12A):1940–56.
59. Walldorf U, Gehring WJ. *empty spiracles*, a gap gene containing a homeobox involved in *Drosophila* head development. EMBO J. 1992;11(6):2247–59.
60. Hartmann B, Hirth F, Walldorf U, Reichert H. Expression, regulation and function of the homeobox gene *empty spiracles* in brain and ventral nerve cord development of *Drosophila*. Mech Dev. 2000;90(2):143–53.
61. Cecchi C, Boncinelli E. *Emx* homeogenes and mouse brain development. Trends Neurosci. 2000;23(8):347–52.
62. Cecchi C. *Emx2*: a gene responsible for cortical development, regionalization and area specification. Gene. 2002;291(1–2):1–9.

63. Simeone A, Gulisano M, Acampora D, Stornaiuolo A, Rambaldi M, Boncinelli E. Two vertebrate homeobox genes related to the *Drosophila empty spiracles* gene are expressed in the embryonic cerebral cortex. EMBO J. 1992;11(7):2541–50.

64. Gulisano M, Broccoli V, Pardini C, Boncinelli E. *Emx1* and *Emx2* show different patterns of expression during proliferation and differentiation of the developing cerebral cortex in the mouse. Eur J Neurosci. 1996;8(5):1037–50.

65. Yoshida M, Suda Y, Matsuo I, Miyamoto N, Takeda N, Kuratani S, et al. *Emx1* and *Emx2* functions in development of dorsal telencephalon. Development. 1997;124(1):101–11.

66. Mallamaci A, Mercurio S, Muzio L, Cecchi C, Pardini CL, Gruss P, et al. The lack of *Emx2* causes impairment of Reelin signaling and defects of neuronal migration in the developing cerebral cortex. J Neurosci. 2000;20(3):1109–18.

67. Galli R, Fiocco R, De Filippis L, Muzio L, Gritti A, Mercurio S, et al. *Emx2* regulates the proliferation of stem cells of the adult mammalian central nervous system. Development. 2002;129(7):1633–44.

68. Gangemi RM, Daga A, Marubbi D, Rosatto N, Capra MC, Corte G. *Emx2* in adult neural precursor cells. Mech Dev. 2001;109(2):323–9.

69. Heins N, Cremisi F, Malatesta P, Gangemi RM, Corte G, Price J, et al. *Emx2* promotes symmetric cell divisions and a multipotential fate in precursors from the cerebral cortex. Mol Cell Neurosci. 2001;18(5):485–502.

70. Adachi Y, Nagao T, Saiga H, Furukubo-Tokunaga K. Cross-phylum regulatory potential of the ascidian *Otx* gene in brain development in *Drosophila melanogaster*. Dev Genes Evol. 2001;211(6):269–80.

71. Acampora D, Pilo Boyl P, Signore M, Martinez-Barbera JP, Ilengo C, Puelles E, et al. OTD/OTX2 functional equivalence depends on 5′ and 3' UTR-mediated control of *Otx2* mRNA for nucleo-cytoplasmic export and epiblast-restricted translation. Development. 2001;128(23):4801–13.

72. McGinnis W, Krumlauf R. Homeobox genes and axial patterning. Cell. 1992;68(2):283–302.

73. Duboule D, Morata G. Colinearity and functional hierarchy among genes of the homeotic complexes. Trends Genet. 1994;10(10):358–64.

74. Akam M. *Hox* and *HOM*: homologous gene clusters in insects and vertebrates. Cell. 1989;57(3):347–9.

75. Trainor PA, Krumlauf R. Patterning the cranial neural crest: hindbrain segmentation and Hox gene plasticity. Nat Rev Neurosci. 2000;1(2):116–24.

76. Studer M, Lumsden A, Ariza-McNaughton L, Bradley A, Krumlauf R. Altered segmental identity and abnormal migration of motor neurons in mice lacking *Hoxb-1*. Nature. 1996;384(6610):630–4.

77. Gavalas A, Studer M, Lumsden A, Rijli FM, Krumlauf R, Chambon P. *Hoxa1* and *Hoxb1* synergize in patterning the hindbrain, cranial nerves and second pharyngeal arch. Development. 1998;125(6):1123–36.

78. Studer M, Gavalas A, Marshall H, Ariza-McNaughton L, Rijli FM, Chambon P, et al. Genetic interactions between *Hoxa1* and *Hoxb1* reveal new roles in regulation of early hindbrain patterning. Development. 1998;125(6):1025–36.

79. Wurst W, Bally-Cuif L. Neural plate patterning: upstream and downstream of the isthmic organizer. Nat Rev Neurosci. 2001;2(2):99–108.

80. Rhinn M, Brand M. The midbrain–hindbrain boundary organizer. Curr Opin Neurobiol. 2001;11(1):34–42.

81. Liu A, Joyner AL. Early anterior/posterior patterning of the midbrain and cerebellum. Annu Rev Neurosci. 2001;24:869–96.

82. Joyner AL, Liu A, Millet S. *Otx2*, *Gbx2* and *Fgf8* interact to position and maintain a mid-hindbrain organizer. Curr Opin Cell Biol. 2000;12(6):736–41.
83. Simeone A. Positioning the isthmic organizer where *Otx2* and *Gbx2* meet. Trends Genet. 2000;16(6):237–40.
84. Hirth F, Kammermeier L, Frei E, Walldorf U, Noll M, Reichert H. An urbilaterian origin of the tripartite brain: developmental genetic insights from *Drosophila*. Development. 2003;130(11):2365–73.
85. Noll M. Evolution and role of Pax genes. Curr Opin Genet Dev. 1993;3(4):595–605.
86. Chiang C, Young KE, Beachy PA. Control of *Drosophila* tracheal branching by the novel homeodomain gene *unplugged*, a regulatory target for genes of the bithorax complex. Development. 1995;121(11):3901–12.
87. Erwin DH, Davidson EH. The last common bilaterian ancestor. Development. 2002;129(13):3021–32.

Early Neurogenesis and Gliogenesis in *Drosophila*

4

Boris Egger

What You Will Learn in This Chapter

In *Drosophila*, neurons and glial cells derive from neural precursor cells called neuroblasts glioblasts or neuroglioblasts. These neural precursor cells reveal many characteristics reminiscent of neural stem cells that we find in the mammalian central nervous system. We will learn how neuroblasts are generated and specified in a spatio-temporal manner in the developing *Drosophila* embryo. The relevance of symmetric versus asymmetric stem cell divisions will be conveyed and the major genetic pathway controlling embryonic neurogenesis will be introduced.

4.1   Introduction

Once the domains of the neuroectoderm are established through inductive interactions and inhibitions, the next important step is to specify groups of cells within the neuroectoderm that are able to diversify and commence to proliferate. This step is carried out primarily by the actions of intrinsic proneural and neurogenic genes expressed within the neuroectoderm. Several basic helix-loop-helix transcription factors, homologs of the *Drosophila acheate-scute* and *atonal* types of proneural genes, are expressed during early neural development [1, 2]. Subsequently, within proneural populations, neural precursor cells are selected for further differentiation. This step involves lateral inhibition mediated

B. Egger (✉)
Department of Biology, University of Fribourg, Fribourg, Switzerland
e-mail: boris.egger@unifr.ch

© Springer Nature Switzerland AG 2023
B. Egger (ed.), *Neurogenetics*, Learning Materials in Biosciences,
https://doi.org/10.1007/978-3-031-07793-7_4

by the interaction of the neurogenic factors Notch and Delta. There are several indications that the principal mechanisms of Notch signalling are conserved from insects to vertebrates [3, 4].

Once neural precursor cells are specified, they proliferate and generate the remarkable cellular diversity that characterizes the central nervous system (CNS), including the brain. Common neural precursor cells in flies and in mice can give rise to both major cell types in the brain, neurons and glia, and the pattern and timing of cell determination is genetically programmed [5, 6]. In *Drosophila,* expression of the transcription factor encoding *glial cells missing* (*gcm*) gene is necessary and sufficient to drive a neural precursor cell into a glial pathway [7]. In vertebrates, neuro-gliogenesis seems to be based on more complex genetic networks [8].

4.2 Embryonic Neurogenesis in the *Drosophila* CNS

The *Drosophila* embryonic brain and ventral nerve cord arise from a sheet of neuroectodermal cells. By interactions among cells that make up proneural equivalence groups of five to six cells, single precursor cells are selected to acquire a neural progenitor or so-called neuroblast fate. In *Drosophila*, the presumptive neuroblast enlarges and delaminates into the interior of the embryo (Fig. 4.1). The remaining cells of each proneural cluster either retain an undifferentiated state or adopt an alternative epidermal fate. Subsequently, to delamination, neuroblasts begin to divide asymmetrically in a stem cell-like manner along the apical-basal axis (Fig. 4.1). In each division, the neuroblast renews itself and buds-off a smaller daughter cell, the ganglion mother cell (GMC). The GMC divides once more to produce two postmitotic neurons that start their terminal differentiation program. Five sequential waves of neuroblast delamination result in the formation of a stereotyped pattern of 30 neuroblasts per hemisegment in the ventral nerve cord. The identity and specification of each neuroblast is determined by its time of birth and position along the body axis [10, 11]. Embryonic neuroblasts can be unambiguously identified based on size, position, and the expression of molecular markers. Toward the end of embryogenesis, most neuroblasts stop dividing and remain quiescent in the neurogenic regions until larval stages. At the end of the first larval instar these same neuroblasts start to proliferate again and give rise to "secondary" neuronal progeny [12, 13]. The neuroblast reactivation and exit of quiescence in the early postembryonic brain are regulated by the uptake of nutrition and the insulin signalling pathway [14, 15].

4.2.1 Neural Competence and Proneural Genes

Genetic studies have provided evidence that a small number of proneural genes, which encode basic helix-loop-helix (bHLH) transcription factors are necessary and sufficient to initiate neural differentiation in the neuroectoderm. Molecular studies identified four

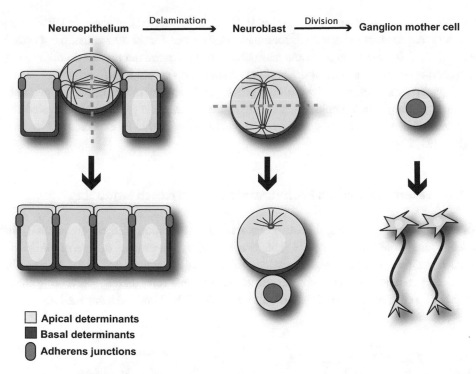

Fig. 4.1 Classical division modes of *Drosophila*n neural progenitors. Most *Drosophila* neuroblasts arise from the embryonic neuroectoderm. Neuroectodermal cells divide symmetrically within the plane of the epithelium generating equal daughter cells. Neuroblasts rotate their spindle axis 90° and undergo asymmetric divisions along the apico-basal axis. Each asymmetric neuroblast division self-renews the mother cell and generates a smaller ganglion mother cell (GMC) that divides once more to give rise to two postmitotic neurons. Adapted from [9]

proneural genes belonging to the *acheate-scute complex* (*asc*), namely *acheate* (*ac*), *scute* (*sc*), *lethal of scute* (*l'sc*), and *asense* (*ase*) [16, 17]. More recently a further proneural gene, *atonal* (*ato*) was isolated, which together with two other *ato* related genes (*amos* and *cato*), comprises the *ato* family [1]. Proteins of the *asc* and *ato* families share features that characterize them as proneural. They are expressed initially in proneural clusters, which are distributed in patterns that foreshadow the distribution of neural progenitor cells in the peripheral and central nervous systems. In addition, they have the ability to trigger the selective generation of neural progenitor cells that probably precedes the process of neuroblast delamination. The most important proneural gene in the embryonic CNS is probably *l'sc,* since its mutational inactivation leads to defective CNS neuroblast formation and since it is the only gene that is restricted to the CNS neuroectoderm [18, 19]. Loss of function analysis showed that *ac* and *sc* are also essential in CNS development and are responsible for the generation of a large subset of neuroblasts. In contrast to the other *asc* genes, the fourth gene of the complex, *ase*, is not expressed in clusters of ectodermal cells, but instead in all progenitors of the PNS and CNS after they have been generated. It appears that *ase* is required for the correct differentiation rather than for the determination of neural progenitors [20]. Interestingly, some neuroblasts in the ventral nerve cord do not require the functional activity of any of the currently known *asc*- or *ato*-related proneural genes. This suggests the presence of additional proneural family members in the fly genome. However, although the complete sequencing of the *Drosophila* genome has revealed the existence of new bHLH genes, none of them shows expression in expected neurogenic regions [21, 22].

4.2.2 Lateral Inhibition: Proneural Genes and Notch Signalling

An essential role of proneural proteins involves the restriction of their own activity in order to single out progenitor cells. This is achieved by the process of *lateral inhibition* and is based on a molecular regulatory loop between adjacent cells (Fig. 4.2). As a result, proneural genes inhibit their own expression in adjacent cells thereby preventing these neighboring cells from adopting a neuroblast fate [23]. The *Notch* pathway mediates these

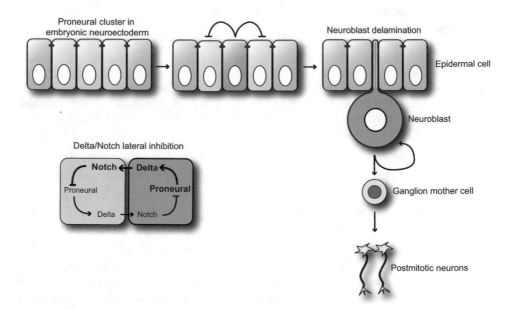

Fig. 4.2 Formation of neuroblasts in the *Drosophila* embryo. Initially, all cells in a proneural cluster are equally competent to become neuroblaststs. Through a process of lateral inhibition mediated by Delta-Notch signalling, a single neuroectodermal cell is determined to become the neuroblast. It delaminates toward the basal side while the remaining neuroectodermal cells adopt an epidermal fate. With each asymmetric division the neuroblast self-renews and produces several ganglion mother cells (GMCs). The ganglion mother cell divides once more to produce two post-mitotic neurons

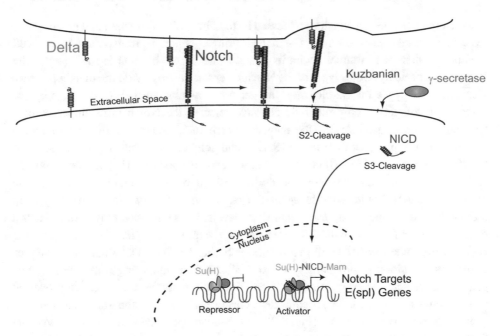

Fig. 4.3 Delta-Notch signalling between two neighboring cells. Notch (red) is transported to the cell membrane where it can interact with its ligand Delta (green). Notch binding to the Delta leads to two proteolytic cleavages of the Notch receptor. In *Drosophila* the metalloprotease Kuzbanian (blue) catalyzes the S2-cleavage and generates a substrate for S3-cleavage by the γ-secretase complex. The proteolytic cleavage releases the Notch intracellular domain (NICD), which enters the nucleus and interacts with the transcription factor Su(H) (green). Upon binding, the repressive Su(H) complex recruits co-factors such as Mastermind (Mam) (orange) and transforms into an activator complex that induces the expression of Notch target genes

events through its ability to repress the expression of the proneural genes (Fig. 4.3). Notch and Delta are transmembrane proteins with extracellular domains partially homologous to vertebrate epidermal growth factor (EGF). These two proteins were shown to form heterophilic interactions and act mainly between neighboring cells. Delta is expressed high in the presumptive neuroblast and signals to Notch receptors expressed in the presumptive epidermal precursors. EGF-repeats and a specific "DSL" (Delta-Serrate-Lin) binding region on the ligand were found in a variety of receptor-ligand systems of diverse functions. The intracellular domain of Dl-like ligands is short, whereas those of N-like receptors are very large and include signals for nuclear transport. In the signal-sending cell, the ligand Delta interacts with the E3-ubiquitin ligase Neuralized (Neur) for ubiquitylation and activation. Delta is then competent to bind to the Notch receptor and induces signalling. Prior to Neur interaction, Delta is inactive and might be endocytosed and degraded.

In the signal-receiving cell, the Notch receptor is produced in the endoplasmatic reticulum (ER) where it interacts with the O-fucosyl transferase (O-Fut) and is transported to the Golgi. In the Golgi Notch is processed by the Furin-like convertase and glycosylated by glycosyltransferases such as O-Fut and Fringe. Notch is transported to the cell membrane where it can interact with its ligand Delta. Notch binding to the ligand Delta leads to two proteolytic cleavages of the Notch receptor. In *Drosophila* the metalloprotease Kuzbanian catalyzes the S2-cleavage and generates a substrate for S3-cleavage by the γ-secretase complex. The proteolytic cleavage releases the Notch intracellular domain (NICD), which translocates to the nucleus. There the intracellular domain of Notch interacts with Suppressor of Hairless [Su(H)] and activates the transcription of genes of the *Enhancer of Split* [*E(spl)*] complex. These genes encode bHLH-type transcriptional repressors that directly downregulate the expression of the proneural genes. As the proneural factors would activate again the expression of *Delta* this inhibitory loop ultimately leads to decreasing *Delta* expression levels. Cells with lower levels of proneural gene activity or *Delta* activity fail to strongly activate Notch signalling in neighboring cells. At the same time, the cell that possesses higher levels of Delta activity continues to mediate inhibition of proneural gene expression in its neighboring cells. These events generate a feedback loop, which quickly amplifies initial subtle differences in proneural gene activity among cells in an equivalent cluster and leads to the stable selection of a single prospective neuroblast (Fig. 4.2). In this developmental context, *Notch* loss-of-function leads to severe neurogenic phenotypes since lateral inhibition is blocked and, thus, the majority of proneural cluster cells delaminate and enter a neural instead of an epidermal pathway [24, 25].

More recently it has been discovered that high Delta levels in the same cell can also bind to Notch and inhibit signalling. This is a mechanism called ligand-mediated *cis*-inhibition. Cell culture experiments and computational models, which integrate trans-activation and *cis*-inhibition led to a better understanding of lateral inhibition. The models can explain how neighboring cells can rapidly become either the signal-sending or the signal-receiving cell [26].

Neurogenic genes such as Notch, Delta, and the Enhancer of Split genes are arranged in an interacting ("epistatic") genetic chain, E(spl) acting most downstream. Epistatic relations can be studied by rescue experiments whereby it is tested whether a particular neurogenic gene can reverse a mutant phenotype of another neurogenic gene. The establishment of an epistatic chain does predict a genetic interaction but not the molecular nature of interactions, i.e., whether they act at the transcriptional, translational, or protein level.

To study which of the neurogenic genes are acting in the signal sending cell (ligand) and which genes are acting in the signal-receiving cell (receptor) marked undifferentiated cells from neurogenic regions of diverse mutants were transplanted to similar wild-type regions. For example, if mutant cells are able to form epidermal cells in wild-type tissue, their receptor for the inhibitory signal must be intact. Hence, the mutant acts on the sender side. As illustrated, it turned out that for example N and E(spl) are on the receptor side (cell autonomous), while other genes such as Delta and Neur have a role in the signal-sending cell (Fig. 4.3).

4.2.3   Spatio-temporal Generation of Neuroblast Progeny

During neural development, neural progenitors generate distinct cell types over time. Populations of such multipotent progenitors are found in the vertebrate cortex, in the retina and the spinal cord. One big question is what are the extrinsic and intrinsic factors that regulate neuronal diversity during development. *Drosophila* neuroblasts produce near invariant lineages, usually consisting of distinct cell types. In addition, neuroblasts generate their progeny cells in a stereotype birth order. Hence, *Drosophila* neuroblasts provide a good model system for studying how the temporal identity of neural precursors is regulated to generate neuronal diversity (Fig. 4.4) [27].

We can think of two ways on how temporal identity is regulated.

(a) Intrinsic regulation of temporal identity: Here, the progenitor acquires initially a unique spatial identity that is based on its anterior-posterior and dorso-ventral position within the developing nervous system. The progenitor then becomes independent on spatial patterning cues and produces a stereotype cell lineage over time. The

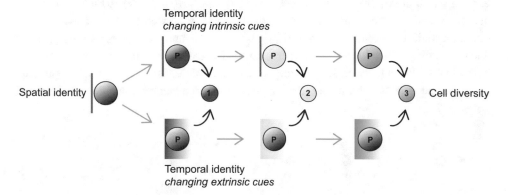

Fig. 4.4 Models for specifying temporal identity. Progenitors (P) generate temporally distinct progeny over time. (Top) Intrinsic temporal identity factors. Progenitors acquire spatial identity (left panel), and then are able to generate the proper sequence of cell types in the absence of extrinsic cues (e.g., *in vitro*). This model requires an intrinsic program that can specify temporal identity but does not rule out extrinsic influences (e.g., feedback inhibition from within a clone). Different progenitors can generate different cell types owing to their initial spatial heterogeneity, which may explain why cell type and birthdate are not correlated in some systems (e.g., neuroblasts and retina). (Bottom) Extrinsic temporal identity factors. Progenitors acquire an initial spatial identity, and then temporal changes in the spatial patterning cues result in the ordered production of different cell types. This model requires a mechanism for temporal regulation of extrinsic cues; it predicts (**a**) that sequential cell-type production is unlikely to occur without changes in extrinsic cues (e.g., single-cell cultures) and (**b**) that there will be a direct link between spatial patterning cues and cell-type specification. Cells undergoing terminal division are likely to be exposed to the same spatial patterning cue environment, which may explain why cell type and birthdate are well correlated in some systems (e.g., cortex, spinal cord, and retina). Redrawn and legend adapted from [27]

composition of the lineage is therefore strictly based on progeny birth order (intrinsic cues) and not affected by environmental changes.

(b) Extrinsic regulation of temporal identity: Here, the progenitor also acquires a spatial identity on the basis of its anterior-posterior and dorso-ventral position within the developing nervous system. However, progenitor identity then changes in response to alteration in the environment (extrinsic cues).

In vitro isolation and transplantation experiments can provide evidence for one or the other model to specify temporal progenitor identity. *Drosophila* neuroblasts faithfully recapitulate the *in vivo* division pattern *in vitro*, and to date only intrinsic factors have been implicated in the regulation of their temporal identity [28]. During the proliferative activity of neuroblasts, the temporal sequence of neuronal progeny generation is reflected in spatially ordered layers. Generally, the first-born, oldest progeny is positioned in deeper layers in the developing CNS, while the last-born, younger cells are found in more superficial layers. Strikingly, following their specification and delamination, most of the neuroblasts go through a characteristic gene expression program during the generation of their progeny (Fig. 4.5) [29, 30]. In an early phase after delamination, the neuroblasts express the transcription factors *hunchback* (*hb*) and *Krüppel* (*Kr*), which seem to be necessary and sufficient for specification of the early generated progeny. Loss of *hb* in the embryonic CNS leads to a loss of first-born GMCs and the corresponding progeny due to cell death, fate skipping, or fate transformation leading to a duplication of the next-born identity. In contrast, continuous misexpression of *hb* leads to supernumerary progeny cells that adopt a first-born fate. Presumptive later-born neurons acquire markers and

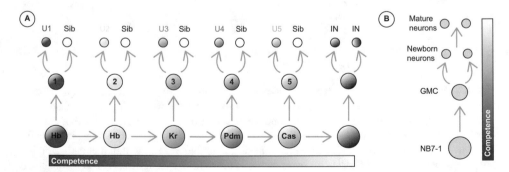

Fig. 4.5 Hb → Kr → Pdm → Cas expression and the competence to respond to Hb in the *Drosophila* NB7–1 lineage. (**a**) Hb, Kr, Pdm, and Cas expression is indicated by color, with transient neuroblast expression maintained by the GMCs and neurons (U1–U5) born during each window of gene expression. In this lineage, two GMCs are born during the window of Hb expression, although most lineages (e.g., NB7-3) produce only one GMC during the Hb expression window. NB7-1 shows progressive loss of competence to generate early-born neurons in response to Hb over time. (**b**) The competence to generate early-born neurons in response to Hb is lost by the time a neuron becomes postmitotic. Redrawn from [27]

morphology of early-born neurons [31, 32]. Similarly, in the absence of *Kr*, second-born cell types are missing and the forced expression of Kr in neuroblasts can again lead to extra cells, which adopt the second-born fate. It was also shown that with time the neuroblast loses competence to generate early-born progeny cells in response to Hb expression. Postmitotic neurons cannot be transformed any longer to an early-born fate. Somewhat later, neuroblast lineages are characterized by the expression of other genes encoding transcription factors in sequential temporal windows namely *pou-homeodomain proteins 1 and 2* (*pmd*), *castor* (*cas*), and possibly *grainyhead* (*grh*). These observations give rise to a model where the temporal expression of *Hb, Kr, Pdm, Cas, Grh* transcription factors specifies the sequentially generated offspring of defined neuroblast lineages. Loss- and gain-of-function experiments suggest extensive cross-regulation among these transcription factors such that the earlier expressed transcription factor activates the next gene in the pathway and concomitantly represses the "next plus one" gene. In addition, it is likely that other inputs participate in the temporal regulation of gene expression in neuroblasts and these are tightly linked with cell cycle progression [29, 32].

4.3   Gliogenesis in *Drosophila*

In addition to neuroblasts that produce neurons, the neuroectodermally derived glial cells in the *Drosophila* embryo can be produced in pure lineages by glioblasts or in mixed lineages by neuroglioblasts (Fig. 4.6) [33–35]. In neuroglioblast lineages, different cell specification processes and division patterns have been described. In some lineages the bifurcation between the neuronal and the glial lineage occurs in the primary division to produce a glioblast and a neuroblast [36, 37]. In other lineages the neuroglioblast first buds-off a neurogenic ganglion mother cell before daughter cells are generated, which give rise to neuron/glia sibling pairs [38]. It is noteworthy that in several lineages neuronal fate precedes glial fate and that tightly regulated timing mechanisms are probably responsible for the generation from one fate to another.

The transcription factor, *glial cells missing/glial cells deficient (gcm/glide)* is necessary and sufficient to induce gliogenesis in all described neuroectodermally derived lineages. In *gcm* mutants, presumptive glial cells are transformed into neurons and, conversely, when *gcm* is ectopically misexpressed, presumptive neurons enter a glial differentiation path [39–41]. Thus, *gcm* seems to be a key regulator in the binary decision between glial versus neuronal development. In the current model, *gcm*-downstream gene activation leads to the promotion of glial differentiation and concomitant repression of neuronal differentiation [42]. Experiments indicate that Notch signalling acts upstream of *gcm* in promoting the gliogenic pathway in the embryonic CNS. In restricted neuroglioblast lineages, Notch signalling is required for glial cell fate specification as the ganglion mother cell divides asymmetrically to generate a neuronal/glial sibling pair. One of the daughter cells inherits the asymmetric factor Numb, which acts to inhibit Notch signalling and this leads to a neuronal cell fate. A clonal analysis of a defined neuroglioblast lineage demonstrated that

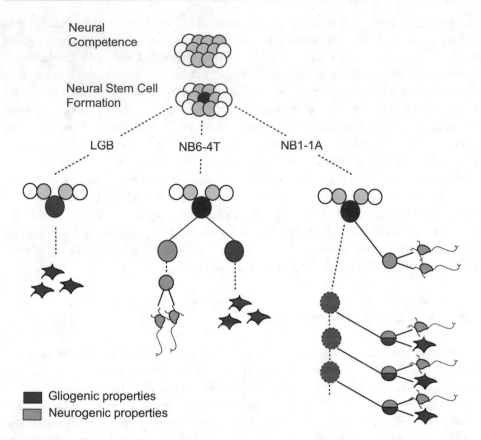

Fig. 4.6 Neurogenic and gliogenic progenitor types. Lateral glioblasts generate gliogenic progeny only. NB6-4T and NB1-1A are neuroglioblasts that produce progeny in mixed lineages. In the NB6-4T lineage the first division leads to the generation of a neuroblast (green) and a glioblast (red). Subsequently, the lineages are separated in neurogenic and gliogenic. In the NB1-1A lineage the first asymmetric division leads to neurogenic ganglion mother cell. However, three subsequent asymmetric divisions each lead to mixed neurogliogenic ganglion mother cells. Each of those ganglion mother cells produces one postmitotic neuron and one postmitotic glial cells

constitutively activated Notch leads to supernumerous glial cells in the progeny, whereas in *Notch* mutants the ganglion mother cells produce two neurons, instead of a neuron/glial sibling pair [38]. Notch promotes glial cell fate through the activation of *gcm*, while Numb acts as its antagonist by inhibiting Notch-mediated *gcm* activation. The action of Notch is variable and cell context dependent. For example, in the peripheral nervous system Notch activation represses glial cell development and *gcm* activation [43]. It seems that Notch signalling can result in different outcomes depending on the type of neural progenitor cells.

4.4   The GAL4-UAS System

In 1993, Brand and Perrimon published a new technology, the GAL4-UAS system that gave another impulse to the use of *Drosophila* as a genetical model (Fig. 4.7) [44]. GAL4 is a yeast transcriptional activator that regulates the expression of genes located downstream of its target sequences, called UAS (for *U*pstream *A*ctivating *S*equences). Neither the GAL4 protein nor the UAS sequences, which are derived from yeast, are normally present in the genome of *Drosophila*. However, Brand and Perrimon showed that when both elements (GAL4 and UAS transgenes) are introduced in the fly genome, a GAL4 protein is synthesized and is able to activate transcription of the genes located downstream of UAS.

The GAL4 system is a *two-part system*: GAL4 and UAS components are introduced separately into the genome using the P-element transposon technology. The GAL4 bearing line is called the *driver* and the line containing the UAS construct is the *responder*.

This opens important prospects:

- Any desired gene added under the control of UAS will be expressed according to the spatio-temporal expression pattern of the GAL4 line.

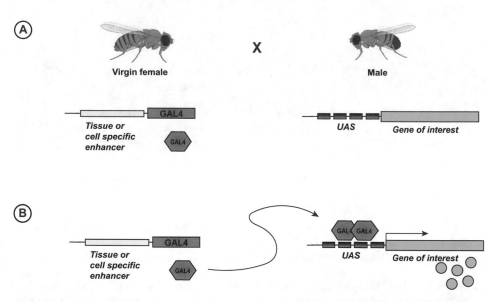

Fig. 4.7 The GAL4-UAS system in *Drosophila*. GAL4 protein is expressed in the driver fly line under a tissue or cell specific enhancer. Without GAL4 in the system, the gene of interest in the responder fly line under UAS control is not expressed. When the driver line is crossed to the responder line the two components are present in the offspring of the cross. The GAL4 protein binds to UAS and induces the expression of the gene of interest in the target cells [44]. In parts created with BioRender.com

- The GAL4-UAS system is widely used to induce cell- or tissue-specific expression transgene constructs. It can be applied to misexpress genes, genetically label cells, or also to knock down gene expression via *UAS-RNAi* constructs.
- Lethal or cytotoxic genes can be kept without harm in a UAS line. Their phenotype will be expressed only when crossed with a GAL4 line and will be restricted to the cells expressing GAL4.
- The GAL4-UAS system has been adapted for several other model organisms such as the nematode *Caenorhabditis elegans* (see Chap. 2) and zebrafish *Danio rerio*.

Question

1. Describe the genetic and cellular process of how neuroblasts are formed in the developing *Drosophila* embryo?
2. What methods would you use to study neuroblast division patterns and neuroblast lineages in the developing brain?

Take-Home Message

- Neuroblasts derive from proneural clusters that contain initially equally competent neural precursor cells.
- Neuroblast determination is mediated through a process of lateral inhibition by the Delta-Notch signalling pathway.
- Neuroblasts change their identity over time as they go through an intrisinsic temporal sequence of transcription factor expression.
- Neural precursor cells are distinguished in neuroblast, glioblasts, and neuroglioblasts.

References

1. Bertrand N, Castro DS, Guillemot F. Proneural genes and the specification of neural cell types. Nat Rev Neurosci. 2002;3(7):517–30.
2. Lee JE. Basic helix-loop-helix genes in neural development. Curr Opin Neurobiol. 1997;7(1):13–20.
3. Beatus P, Lendahl U. *Notch* and neurogenesis. J Neurosci Res. 1998;54(2):125–36.
4. de la Pompa JL, Wakeham A, Correia KM, Samper E, Brown S, Aguilera RJ, et al. Conservation of the Notch signalling pathway in mammalian neurogenesis. Development. 1997;124(6):1139–48.
5. Jones BW. Glial cell development in the *Drosophila* embryo. BioEssays. 2001;23(10):877–87.
6. Temple S, Qian X. Vertebrate neural progenitor cells: subtypes and regulation. Curr Opin Neurobiol. 1996;6(1):11–7.

7. Wegner M, Riethmacher D. Chronicles of a switch hunt: *gcm* genes in development. Trends Genet. 2001;17(5):286–90.

8. Vetter M. A turn of the helix: preventing the glial fate. Neuron. 2001;29(3):559–62.

9. Wu PS, Egger B, Brand AH. Asymmetric stem cell division: lessons from *Drosophila*. Semin Cell Dev Biol. 2008;19(3):283–93.

10. Skeath JB, Thor S. Genetic control of *Drosophila* nerve cord development. Curr Opin Neurobiol. 2003;13(1):8–15.

11. Doe CQ, Fuerstenberg S, Peng CY. Neural stem cells: from fly to vertebrates. J Neurobiol. 1998;36(2):111–27.

12. Truman JW, Bate M. Spatial and temporal patterns of neurogenesis in the central nervous system of *Drosophila melanogaster*. Dev Biol. 1988;125(1):145–57.

13. Datta S. Control of proliferation activation in quiescent neuroblasts of the *Drosophila* central nervous system. Development. 1995;121(4):1173–82.

14. Chell JM, Brand AH. Nutrition-responsive glia control exit of neural stem cells from quiescence. Cell. 2010;143(7):1161–73.

15. Sousa-Nunes R, Yee LL, Gould AP. Fat cells reactivate quiescent neuroblasts via TOR and glial insulin relays in *Drosophila*. Nature. 2011;471(7339):508–12.

16. Ghysen A, Dambly-Chaudiere C. Genesis of the *Drosophila* peripheral nervous system. Trends Genet. 1989;5(8):251–5.

17. Campuzano S, Modolell J. Patterning of the *Drosophila* nervous system: the *achaete-scute* gene complex. Trends Genet. 1992;8(6):202–8.

18. Jimenez F, Campos-Ortega JA. Defective neuroblast commitment in mutants of the *achaete-scute* complex and adjacent genes of *D. melanogaster*. Neuron. 1990;5(1):81–9.

19. Martin-Bermudo MD, Gonzalez F, Dominguez M, Rodriguez I, Ruiz-Gomez M, Romani S, et al. Molecular characterization of the *lethal of scute* genetic function. Development. 1993;118(3):1003–12.

20. Brand M, Jarman AP, Jan LY, Jan YN. *asense* is a *Drosophila* neural precursor gene and is capable of initiating sense organ formation. Development. 1993;119(1):1–17.

21. Moore AW, Barbel S, Jan LY, Jan YN. A genomewide survey of basic helix-loop-helix factors in *Drosophila*. Proc Natl Acad Sci USA. 2000;97(19):10436–41.

22. Peyrefitte S, Kahn D, Haenlin M. New members of the *Drosophila* Myc transcription factor subfamily revealed by a genome-wide examination for basic helix-loop-helix genes. Mech Dev. 2001;104(1–2):99–104.

23. Skeath JB, Carroll SB. The *achaete-scute* complex: generation of cellular pattern and fate within the *Drosophila* nervous system. FASEB J. 1994;8(10):714–21.

24. Artavanis-Tsakonas S, Rand MD, Lake RJ. Notch signaling: cell fate control and signal integration in development. Science. 1999;284(5415):770–6.

25. Bray S. Notch signalling in *Drosophila*: three ways to use a pathway. Semin Cell Dev Biol. 1998;9(6):591–7.

26. Yaron A, Sprinzak D. The cis side of juxtacrine signaling: a new role in the development of the nervous system. Trends Neurosci. 2012;35(4):230–9.

27. Pearson BJ, Doe CQ. Specification of temporal identity in the developing nervous system. Annu Rev Cell Dev Biol. 2004;20:619–47.

28. Homem CC, Reichardt I, Berger C, Lendl T, Knoblich JA. Long-term live cell imaging and automated 4D analysis of *Drosophila* neuroblast lineages. PLoS One. 2013;8(11):e79588.

29. Brody T, Odenwald WF. Cellular diversity in the developing nervous system: a temporal view from *Drosophila*. Development. 2002;129(16):3763–70.

30. Pearson BJ, Doe CQ. Regulation of neuroblast competence in *Drosophila*. Nature. 2003;425(6958):624–8.

31. Isshiki T, Pearson B, Holbrook S, Doe CQ. *Drosophila* neuroblasts sequentially express transcription factors which specify the temporal identity of their neuronal progeny. Cell. 2001;106(4):511–21.

32. Novotny T, Eiselt R, Urban J. Hunchback is required for the specification of the early sublineage of neuroblast 7-3 in the *Drosophila* central nervous system. Development. 2002;129(4):1027–36.

33. Schmidt H, Rickert C, Bossing T, Vef O, Urban J, Technau GM. The embryonic central nervous system lineages of *Drosophila melanogaster*. II. Neuroblast lineages derived from the dorsal part of the neuroectoderm. Dev Biol. 1997;189(2):186–204.

34. Schmid A, Chiba A, Doe CQ. Clonal analysis of *Drosophila* embryonic neuroblasts: neural cell types, axon projections and muscle targets. Development. 1999;126(21):4653–89.

35. Bossing T, Udolph G, Doe CQ, Technau GM. The embryonic central nervous system lineages of *Drosophila melanogaster*. I. Neuroblast lineages derived from the ventral half of the neuroectoderm. Dev Biol. 1996;179(1):41–64.

36. Bernardoni R, Kammerer M, Vonesch JL, Giangrande A. Gliogenesis depends on *glide/gcm* through asymmetric division of neuroglioblasts. Dev Biol. 1999;216(1):265–75.

37. Akiyama-Oda Y, Hosoya T, Hotta Y. Asymmetric cell division of thoracic neuroblast 6-4 to bifurcate glial and neuronal lineage in *Drosophila*. Development. 1999;126(9):1967–74.

38. Udolph G, Rath P, Chia W. A requirement for *Notch* in the genesis of a subset of glial cells in the *Drosophila* embryonic central nervous system which arise through asymmetric divisions. Development. 2001;128(8):1457–66.

39. Vincent S, Vonesch JL, Giangrande A. Glide directs glial fate commitment and cell fate switch between neurones and glia. Development. 1996;122(1):131–9.

40. Jones BW, Fetter RD, Tear G, Goodman CS. *Glial cells missing*: a genetic switch that controls glial versus neuronal fate. Cell. 1995;82(6):1013–23.

41. Hosoya T, Takizawa K, Nitta K, Hotta Y. *Glial cells missing*: a binary switch between neuronal and glial determination in *Drosophila*. Cell. 1995;82(6):1025–36.

42. Giesen K, Hummel T, Stollewerk A, Harrison S, Travers A, Klämbt C. Glial development in the *Drosophila* CNS requires concomitant activation of glial and repression of neuronal differentiation genes. Development. 1997;124(12):2307–16.

43. Van De Bor V, Giangrande A. Notch signaling represses the glial fate in fly PNS. Development. 2001;128(8):1381–90.

44. Brand AH, Perrimon N. Targeted gene expression as a means of altering cell fates and generating dominant phenotypes. Development. 1993;118(2):401–15.

Neural Stem Cells and Brain Tumour Models in *Drosophila*

5

Boris Egger

What You Will Learn in This Chapter

During development organs and tissues grow in size. This growth can be achieved either by increasing the size of individual cells or by increasing the number of cells by cell proliferation. In an organ like the brain, it is crucial to control the number of cells in order to build functional neuronal circuits. In this chapter, we will look at different types of neural precursor cells in the growing *Drosophila* larval brain. Each type of neural precursor cell type in *Drosophila* displays a rather stereotype division pattern. Different modes of division serve to either expand or to differentiate the precursor cell pool. We will learn that different division modes will lead to vastly different outcomes in terms of the generated cell lineage size. The different neural precursor types use also different timings and cellular mechanisms to terminate proliferation once the correct number of progeny cells is reached.

After we learned how neural precursor cells behave under normal conditions, we will explore what can go wrong during neural proliferation as a result of genetic abnormalities. The *Drosophila* brain is an excellent model to study mechanisms of tumour formation as caused by abnormal neural precursor division. We will introduce more genetic tools for *Drosophila* research that allow to genetically induce tumour-inducing mutation in a controlled timely and lineage-restricted manner.

(continued)

B. Egger (✉)

Department of Biology, University of Fribourg, Fribourg, Switzerland

e-mail: boris.egger@unifr.ch

© Springer Nature Switzerland AG 2023

B. Egger (ed.), *Neurogenetics*, Learning Materials in Biosciences,

https://doi.org/10.1007/978-3-031-07793-7_5

The deep molecular and cellular understanding of neural precursor biology that *Drosophila* scientists have collected over several decades of research is quite unique. It turns out that many of the mechanisms that regulate neural precursor proliferation and lineage size are also used by neural stem cells in mammalian species.

5.1   Introduction

A prominent theory in cancer biology proposes that some human cancers originate from only a small group of cancer stem cells that are able to sustain tumour growth. Interestingly, research in *Drosophila* suggested a connection between stem cells and cancers already 30 years ago [1]. The cellular and molecular characteristics of cancer stem cells in humans remain largely elusive. Hence, model organisms such as *Drosophila* serve to address the question of how normal stem cells might be transformed to tumourigenic stem cells that overproliferate and form malignant metastases [2, 3]. Although it is not clear whether tumours occur naturally in *Drosophila*, the loss- or gain-of-function of a number of genes results in a range of tumours from benign hyperplasia to malignant neoplasms that are invasive and lethal to the host. The identified genes fall into two main categories: oncogenes and tumour suppressor genes. Oncogenes have the potential to transform normal cells into tumour cells. In tumour cells, oncogenic mutations often lead to abnormally high expression levels or to overactivated proteins. Tumour suppressor genes protect normal cells from adopting tumourigenic fates. In tumour cells, tumour suppressor genes might be partially or completely inactivated, which leads to an increased likelihood of tumour formation.

The hallmark of human cancer cells can be subdivided into biological capabilities that are acquired during the multistep development of tumour formation [4]. *Drosophila* tumour cells reproduce many hallmarks of human cancer cells and can therefore contribute to the understanding of the mechanisms of cancer formation and progression. However, to gain insights into the biology of cancer stem cells we need to have a good knowledge of how stem cells behave during normal development. In the following sub-chapters, we will look at how neural stem cells divide and proliferate in the developing *Drosophila* brain.

5.2   Symmetric and Asymmetric Stem Cell Divisions

Stem cells divide either symmetrically, producing two identical daughter cells, or asymmetrically producing two different daughter cells. Stem cells may also undergo symmetric differentiative divisions, depleting the stem cell pool and producing two developmentally restricted precursors or post-mitotic progeny (Fig. 5.1). Proliferative symmetric divisions serve to rapidly expand a progenitor pool, whereas asymmetric divisions are important for

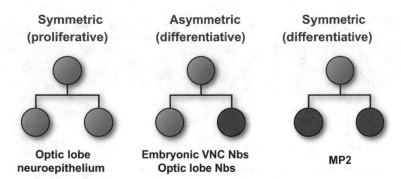

Fig. 5.1 Neural stem cell divisions. Neural stem cells undergo symmetric proliferative or asymmetric differentiative self-renewing divisions. Neural stem cells may also undergo symmetric differentiative divisions, thereby depleting the stem cell pool and producing two developmentally restricted precursors of post-mitotic progeny. Given are some examples for each division type in *Drosophila*. *VNC* ventral nerve cord, *Nbs* neuroblasts, *MP2* Midline precursor 2. Adapted from [5]

differentiation, as they generate daughter cells that can differ in size, mitotic potential, and lineage commitment. In vertebrates, neural stem cells such as neuroepithelial cells in the ventricular zone of the cerebral cortex initially undergo several rounds of proliferative symmetric division to increase their number. Later during neurogenesis, asymmetric divisions are observed when radial glial cells directly generate neurons and self-renew.

In *Drosophila*, neuroectodermal cells in the embryo or neuroepithelial cells in the developing optic lobe undergo symmetric divisions. Both epithelia give rise to neuroblasts, which switch to an asymmetric division mode thereby self-renewing and generating smaller ganglion mother cells (GMCs). One genetic pathway that is important for regulating the switch from symmetric to asymmetric division is the Notch signalling pathway. In vertebrates as well as in *Drosophila* active Notch signalling maintains the proliferative neural stem cell state and prevents the premature switch to a differentiative division mode [6].

5.3   Polarity Cues Direct the Asymmetric Segregation of Cell-Fate Determinants

Epithelial cells and neuroblasts are highly polarised cell types. One key mechanism by which a cell can produce two different daughter cells is by partitioning cell-fate determinants unequally between the daughter cells. Experiments performed on neuroblasts in culture suggest that intrinsic factors are the major players directing a neuroblast's ability to divide asymmetrically and to self-renew. Most of the proteins that direct neuroblast polarity are expressed and localised already in the neuroectodermal cells, where they establish epithelial apico-basal polarity. The neuroblast inherits a complement of these proteins prior to delamination (Fig. 5.2) [5].

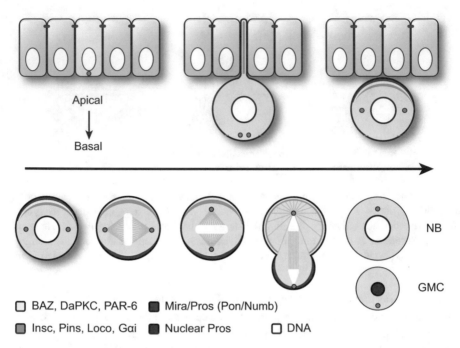

Fig. 5.2 Asymmetric neuroblast division. The subcellular localisation of several polarity proteins, cell-fate determinants and their adaptor proteins are indicated in different colours. In the neuroecto-dermal cells, Baz, DmPar6 and DaPKC localise apically (yellow) as the neuroblast delaminates from the neuroectoderm, Insc is expressed and recruits Pins and Gαi to the apical cortex. The adaptor protein Mira becomes localised to the basal cortex, where it anchors the GMC determinants Pros and Brat. After segregating to the GMC, Mira is degraded and releases Pros, which then enters the GMC nucleus. Adapted from [5]

Embryonic neuroectodermal cells or optic lobe neuroepithelial cells undergo horizontally oriented divisions, within the epithelial plane, resulting in the symmetric distribution of apical and basal cell fate determinants. The neuroblast rotates its mitotic spindle 90° so that subsequent divisions are oriented along the apico-basal axis of the tissue [7, 8]. Spindle orientation and the subsequent asymmetric segregation of cell fate determinants to the basal GMC are dependent on a molecular complex at the apical cortex of the neuroblast [3, 9]. The apical complex consists of the evolutionary conserved Par proteins, Bazooka (Baz, Par3), DmPar6 and the *Drosophila* homologue of the atypical protein kinase C, DaPKC (Fig. 5.3). During neuroblast delamination the adaptor protein Inscuteable (Insc) becomes expressed and binds the Par complex. Insc then recruits the Pins/Gαi complex to the apical side, which is involved in spindle orientation and together with heterotrimeric G proteins in spindle asymmetry. Spindle asymmetry is important to generate the size difference between neuroblast and the ganglion mother cell: the neuroblast has a diameter of 10–12 μm whereas the ganglion mother cell has a diameter of about 4–6 μm. The apical complex directs cell fate determinants to the basal side of the neuroblast cortex. The

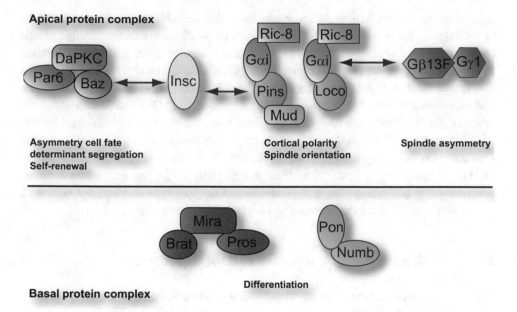

Fig. 5.3 Apical and basal complexes involved in asymmetric neuroblast divisions. Apical and basal protein complexes are illustrated together with their main functions in neuroblast division. The apical Par complex (Par6/Baz/aPKC) is mainly involved in establishing polarity and in directing cell fate determinants to the basal neuroblast cortex. Insc/Pins/Gαi is crucial for coordinating the spindle axis with tissue polarity, whereas free Gβγ affects spindle geometry. At the basal side, cell fate determinants Mira/Pros/Brat and Pon/Numb control proliferation and differentiation in the presumptive GMC. Adapted from [5]

basal determinant Miranda (Mira), an adaptor protein, binds to the homeodomain protein Prospero (Pros), and the NHL domain protein Brain tumour (Brat), and targets them to the basal cortex (Fig. 5.3). After segregation to the basal ganglion mother cell Mira is degraded and Pros is released from the cell cortex to the nucleus where it represses neuroblasts genes and activates differentiation genes [10].

5.3.1   Mechanisms to Orient the Cell Division Axis

In mitotic cells, the orientation of the mitotic spindle dictates the orientation of the cleavage plane and the position of the two daughter cells. As discussed above when progenitor cells are polarised, i.e., along the apical-basal axis, the spindle orientation has an important impact on the distribution of cell fate determinants to the two daughter cells. An intriguing question is how neuroblasts are able to orient their mitotic spindle repeatedly in consecutive divisions and thus their division along the apico-basal axis? What are the cues required to position the Par/Insc/Pins/Gαi complex at the apical cortex? In an experiment with cultured neuroblasts, it was found that neuroblasts grown in contact with epithelial cells are able to maintain the orientation of the division axis for multiple rounds of divisions. In contrast, single cultured neuroblasts are unable to maintain their orientation of division and produce ganglion mother cells (GMCs) in apparently random directions. The results indicate that extracellular signals from the overlying epithelium could have a role in the orientation of neuroblast division (spindle axis) [11]. Possible candidates are among signalling molecules such as extracellular matrix proteins or ligands expressed by neuroectodermal cells. Indeed, the G protein-coupled receptor Tre1 was reported to act via Goα activation upon a presumptive extrinsic signal. Tre1 recruits Pins and subsequently the Par complex to the apical neuroblast cortex [12]. Hence the interaction between neuroectoderm and neuroblasts is a good example on how cells can interact with each other to provide positional information and to establish tissue architecture.

5.3.2   Proliferation and Termination of Neural Precursors

The adult *Drosophila* brain contains about 100,000 neurons. Most of these neurons are generated during a second neurogenesis phase in the growing *Drosophila* larvae and only fully differentiate in the pupal case. The established neuronal circuits will allow the larvae to navigate, forage and mate. In the larval brain, about 100 neuroblast can be identified by the expression of the Hes family transcription factor Deadpan (Dpn) and other neuroblast-specific markers. The neuroblasts found in the developing larvae terminate cell divisions during larval and pupal stages, however, the timing and mechanisms of termination varies between cell lineages located in different regions of the brain. The developing *Drosophila* CNS can be subdivided into the optic lobes, the central brain and the ventral nerve cord. In each of these compartments a stereotype pattern of neuroblasts was described (Fig. 5.4). For instance, in the central brain a defined set of neuroblasts generate the neurons for the olfactory learning and memory centres known as the mushroom bodies. In the ventral nerve cord neuroblasts generate progeny cells that differentiate into inter- and motoneurons. Five major types of neuroblast are well described in the postembryonic brain and can be distinguished based on their mode of division and the mechanism of termination [13, 14]. As discussed previously for embryonic *Drosophila* neuroblasts, also larval neuroblasts go through a temporal cascade whereby a distinct set of transcription factors are sequentially expressed to define neuronal and glial subtypes. Once this transcriptional programme and the lineage composition is completed neuroblasts terminate their proliferative capacity [15].

5.3.2.1 Type IA and Type ID Neuroblasts
Type I neuroblasts have their origin in the embryo where they generate neuronal lineage for the larval nervous system (Figs. 5.4 and 5.5). At the end of embryonic development, many neuroblast undergo apoptosis while a number of selected neuroblasts undergo a transient quiescent phase. They are reactivated during early larval stages to generate neural lineages

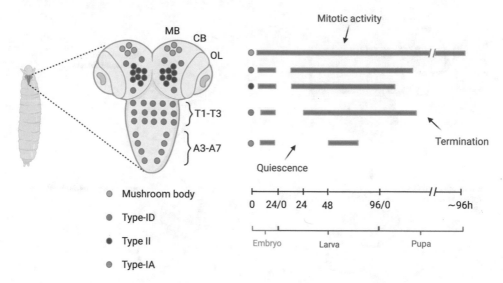

Fig. 5.4 Types of neuroblasts in the developing *Drosophila* larval brain. The central brain (CB) contains Type-ID neuroblasts, four mushroom body (MB) neuroblasts and eight Type-II neuroblasts. The ventral nerve cord (VNC) is subdivided into thoracic neuromeres (T1–T3) and abdominal neuromeres (A1–A7) containing Type-ID and Type-IA neuroblasts, respectively. Each neuroblast lineage is shaped by the length of the mitotic phase, i.e., the timing of reactivation from the quiescence phase and the timing of termination of the mitotic activity. *OL* Optic lobe. Redrawn after [13], created with BioRender.com

for the adult nervous system [16, 17]. Type-IA and Type-ID neuroblasts reveal differences in the mechanisms in how to terminate division after completion of their lineages. Type-IA neuroblasts are found in the abdominal segments and contribute only small lineages of 4–12 neurons to the adult CNS. They are terminated early during late embryonic or larval stages by a mechanism of Hox factor-induced apoptosis [18].

Type-ID neuroblasts are found in the thoracic segments and in the central brain. These neuroblasts are not terminated until pupal stages and thus generate larger lineages of about 100 progeny cells. During pupal stages, cell cycle exit is preceded by a reduction in neuroblast cell size. In other words, neuroblasts do not regrowth any longer after each division as they normally do during larval stages. The reduction in cell size is regulated with a change in energy metabolism. Oxidative phosphorylation and the mitochondrial respiratory chain are required to uncouple the cell cycle from cell growth. Neuroblasts progressively shrink in size with each division. For termination, Prospero (Pros) then enters the nucleus of neuroblast to induce a final differentiative symmetric division [19, 20].

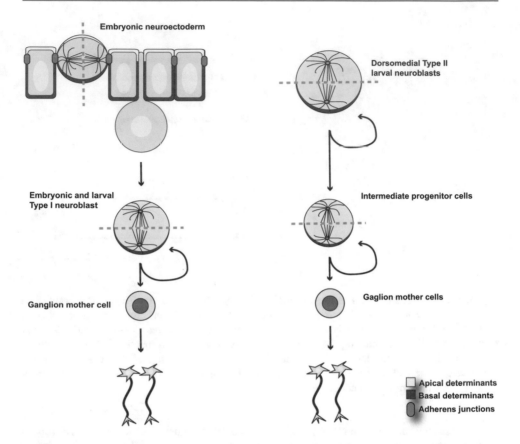

Fig. 5.5 Type I and Type II neuroblast lineages. Type I neuroblasts delaminate from the embryonic neuroectoderm and undergo asymmetric divisions. Thereby the mother cell is self-renewed, and a smaller ganglion mother cell (GMC) is generated. The GMC divides once more to produce two differentiating neurons. Type II neuroblasts undergo self-renewing asymmetric divisions to generate another self-renewing progenitor cell type. These intermediate progenitor cells (INPs) mature prior to dividing asymmetrically and producing GMCs

5.3.2.2 Type II Neuroblasts

Type II neuroblasts are characterised by the absence of the expression of bHLH transcription factor Asense (Ase). This type of neuroblast uses a modified pattern of asymmetric divisions that includes a second type of self-renewing progenitor (Figs. 5.4 and 5.5). In the dorso-medial part of the larval central brain, eight Type II neuroblasts divide asymmetrically to self-renew and generate intermediate neural progenitors (INPs). INPs initially mature over a period of four to six hours while sequentially express Asense (Ase) and Deadpan (Dpn). After the maturation phase INPs undergo three to five additional rounds of asymmetric division, generating another INP and a GMC that divides terminally into two neurons or glial cells. Since each Type II neuroblasts generates several intermediate progenitors, their lineages are up to four-fold larger than Type I neuroblast lineages, with up to 400 and more progeny cells. INPs do not express nuclear Pros and therefore seem to be more susceptible to tumour formation, as we will discuss later. Active Notch signalling is a crucial regulator of Type II specific genetic network. It promotes neuroblast growth and proliferation. Notch also regulates the regrowth of neuroblasts after division. In the more differentiated daughter cells, such as the GMCs, the self-renewing and cell growth network is turned off by Numb and Brat [21–23].

5.3.2.3 Mushroom Body Neuroblasts

Four Mushroom Body neuroblasts (MB) per brain lobe generate the neuronal lineages involved in memory formation and olfactory learning. They are exceptional because these neuroblasts do not enter a quiescent phase at the end of embryogenesis and terminate their division about two days after other brain neuroblasts stopped proliferation (Fig. 5.4). As a consequence, MB neuroblasts generate fivefold more neurons than most Type-ID neuroblasts, up to 500 neurons. These neuroblasts are also atypical in that they do not express or require Pros for termination. Instead, both apoptosis and autophagy are required for the elimination of mushroom body neuroblasts and termination of neurogenesis [24].

5.3.2.4 Optic Lobe Neural Precursors

The most dramatic cell proliferation within the postembryonic CNS occurs in the developing visual anlagen, the so-called optic lobes. The optic lobe proliferation centres derive from an optic placode of embryonic origin [25]. Neurogenesis in the optic lobe involves an early neuroepithelial cell expansion phase via symmetric divisions followed by a late larval differentiation phase involving asymmetric neuroblast divisions. Optic lobe neuroblasts are generated at the medial edge of the neuroepithelium of the outer proliferation centre. A wave of *lethal of scute* (*l'sc*) proneural gene expression precedes the conversion of neuroepithelial cells to neuroblasts (Fig. 5.6). The neuroepithelial-to-neuroblast conversion continues for approximately four days until the neuroepithelium is fully depleted [26–29].

5.3.3 Techniques to Study Tumour Suppressor Gene and Oncogene Function in Genetic Mosaics

Genetic mosaic techniques allow one to induce genetic changes in subsets of cells in an individual organism. Therefore, these techniques permit to study:

- The function of a gene whose mutation is leading to organismic lethality.
- The function of a pleiotropic gene in a specific process, in a specific tissue and, at specific stages of development.
- The development at the level of individual cells and cell lineages.

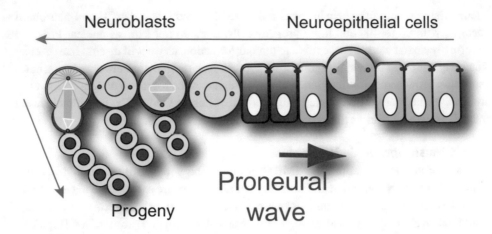

Fig. 5.6 Neuroepithelial to neuroblast transition in the optic lobe. In the outer proliferation centre (OPC) of the optic lobe neuroepithelial cells proliferate through symmetric divisions. At the medial edge of the OPC neuroepithelial cells transform into medulla neuroblasts. The transition is induced by a proneural wave that sweeps across the neuroepithelium. The neuroepithelial cells located in front of the wave expand through symmetric division, while the neuroblasts behind the wave self-renew and generate GMCs through asymmetric divisions

In addition, the juxtaposition of mutant and wild-type clones allows one to test the 'cell-autonomy' of a mutant phenotype. The mosaic techniques described in this chapter were created through the possibility of inserting DNA constructs in the fly genome by applying the P-element technology. They use variations of two different systems, both deriving from yeast: the FLP-FRT system and the GAL4-UAS system.

The FLP-FRT system involves a targeted DNA mitotic recombination. The yeast recombinase Flippase (FLP) drives recombination between its target sequences, the FRTs (for FLP recombination targets). FRTs were inserted into proximal locations on each of the chromosome arms and stocks were generated that express the Flippase under the control of the heat shock promoter hsp-70 (hs-FLP). If a fly has two FRTs at identical positions on homologous chromosomes, heat shock-induced expression of the FLP can cause recombination between the FRT sites.

Since the FLP recombinase is induced by heat shock, one can control:

- The clone size: if mitotic recombination is induced in a precursor cell early in a cell lineage, half of its progeny will be homozygous mutant. For example, if recombination occurs in the fertilised egg, half of the organism will be homozygous mutant while the other half will be wild-type. On the other extreme, if recombination is induced in a cell that will only divide once, a single-cell mutant cell and its wild-type sister cell will be observed.
- The clone frequency: increased intensity of the heat shock by higher temperature or prolonged heat shock leads to an increase in the synthesis of FLP recombinase.

Therefore, the probability of recombination is increased in each pre-mitotic cell leading to increased numbers of clones.

5.3.4   Mosaic Analysis with a Repressible Cell Marker (MARCM)

The MARCM technique (Mosaic analysis with a repressible cell marker) involves the GAL4 inhibitor GAL80 (Fig. 5.7) [31]. The principle of this technique is to generate and to visualise at the same time mutant or control clones using the GAL4-UAS system. For this purpose, flies are generated that are heterozygous for a given mutation *m*, i.e., in a tumour suppressor gene. The flies also carry a *GAL4* driver construct:

- Carry a *tubulin-GAL80* construct on the homologous chromosome, for ubiquitous GAL80 expression
- Chromosome arms with the mutation (*m*) or GAL4 and the tub-GAL80 construct also carry a proximal FRT site
- Contain a *UAS*-reporter on other chromosomal positions and a *hs*-FLP transgene (often on the X chromosome)

Without heat shock, the entire animal is heterozygous for *m* or *GAL4*, as well as for *tub-GAL80*. Therefore, the mutation has no effect. GAL4 function is inhibited by the GAL80 repressor and cannot drive the expression of the *UAS*-reporter gene. When a heat shock is applied, the FLP is expressed and induces recombination between the FRTs in mitotically active cells in a stochastic manner. In cases where recombination occurred daughter cells either become homozygous for *m/GAL4* or for *tub-GAL80*. Since the cells homozygous for *GAL4* do not express GAL80 any longer, expression of GAL4 and hence the UAS-driven reporter gene is induced. The homozygous mutant cells are therefore the only cells that are labelled by the expression of the reporter. Since the recombination process depends on mitosis the expression of the *hs*-FLP construct needs to be induced prior to the last division of the target cell lineage.

5.3.5   Mosaic Analysis with Flip-Out Clones

The FLP-FRT technology can be used to remove stretches of DNA in *cis* that are located between two tandem FRT sequences with the same directionality (Fig. 5.8).

This method can be used for inducing misexpression of selected genes, i.e., oncogenes in cell clones. It is generally coupled to the GAL4-UAS system. For this purpose, an 'FLP-out cassette' containing two FRTs flanking a DNA sequence that includes a STOP codon can be inserted between *UAS* and a reporter gene. Hence, flies bear the following constructs:

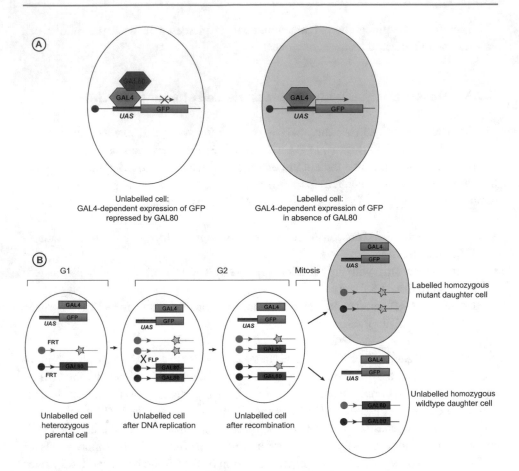

Fig. 5.7 Mosaic analysis with a repressible cell marker (MARCM). The MARCM system enables the generation of homozygous mutant cellular clones in an otherwise wild-type or heterozygous background. (**a**) In the parental cells, GAL4 activity is blocked by the repressor GAL80, which is usually expressed ubiquitously under by a tubulin promoter. GFP under the control of UAS is not expressed (unlabelled cell). Loss of GAL80 will allow the expression of GFP (labelled cell). (**b**) In G2 phase of the cell cycle FLP mediates the exchange of non-sister chromatids between the homologous chromosomes, distal to the FRT sites. The segregation of chromosomes after the recombination event leads upon mitosis to two different daughter cells. The GFP labelled cell is homozygous for the mutation (yellow star) while the other cell which carries two wild-type allele stays unlabelled due to the inheritance of the GAL80 repressor constructs. Redrawn after [30]

- The *UAS*–FLP-out cassette–reporter gene
- A *GAL4* driver construct
- *hs*-FLP allowing heat shock-dependent synthesis of FLP recombinase

The chromosomal locations of the different constructs are not important in this case.

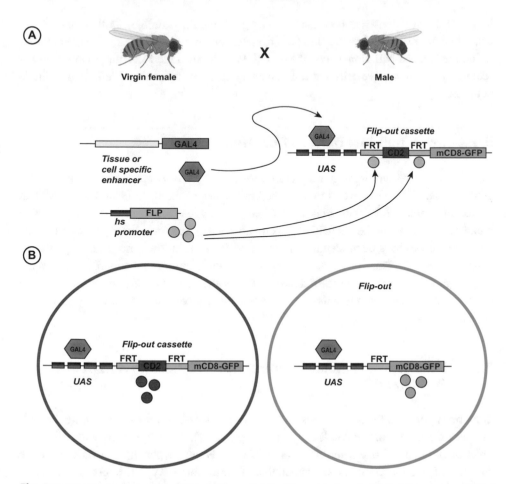

Fig. 5.8 Mosaic analysis with flip-out clones. (**a**) In this example, virgin females carry a GAL4 transgene under the control of a tissue-specific enhancer and a Flippase transgene under the control of a heat shock promoter. The males carry a flip-out construct whereby an FRT > CD2 > FRT stop cassette is inserted in between the Upstream Activation Sequence (UAS) and the mCD8-GFP gene. In the progeny of the fly cross all constructs are present in the cells and GAL4 binds the UAS. (**b**) When the stop cassette is in place GAL4-UAS drives the expression of CD2, which can serve as a marker to label the entire target tissues (red). Upon a heat shock the Flippase is induced and catalyzes recombination at FRT target sites. This leads to the deletion or 'flip-out' of the stop cassette in a stochastic manner. In such cases, GAL4-UAS drives the expression of mCD8-GFP, which labels the membranes of clonal cells (green). In parts created with BioRender.com

Due to the presence of the STOP codon between the coding and the promotor sequences, the reporter gene cannot be expressed despite the presence of GAL4. A heat shock inducing the synthesis of FLP recombinase leads to the excision of the 'FLP-out cassette' in some cells in stochastic manner. In these cells only, the reporter gene will be expressed under the control of *UAS*. Alternatively, the 'FLP-out cassette' can be inserted in a GAL4 driver construct between the *Gal4* gene and the promotor. In this case, GAL4 cannot be expressed unless the 'FLP-out cassette' is removed. In contrast to the methods to induce labelled cell clones, i.e., the MARCM technique, the FLP-out system does not require mitotically active cells and can also be induced in differentiated cell, such as single neurons.

5.4   Brain Tumours Derived from Neuroblasts

The first tumour suppressor genes were discovered in *Drosophila* as mutations in these genes caused a massive overgrowth of the brain lobes in the third instar larvae. Mutant alleles of tumour suppressor genes are often homozygous lethal at early stages hence mutant clonal systems (i.e., MARCM system) are used to study tumour suppressor activity. Dozens of genes have been identified in *Drosophila* that fulfil the function of suppressing neoplastic growth. The tumours caused by mutation in these tumour suppressor genes are in general not invasive and proliferation ceases as they undergo terminal differentiation concomitant with surrounding tissue. Hence, the tumour cells do not show any metastatic behaviour [32].

5.4.1   Examples of Tumour Suppressor Genes Affecting Neuroblast Proliferation

Tumour suppressor function is best studied in the larval brain by generating loss-of-function clones by the MARCM system (Fig. 5.7). Neuroblast clones with disrupted function of basal cell fate determinants Pros, Brat or Numb display neoplastic overgrowth and lack of differentiation. It is assumed that in these clones, Type II intermediate neural progenitors and GMCs fail to mature, adopt a neuroblast-like fate and continue to divide in a proliferative mode [33].

Allograft cultures are another approach to study the malignancy of mutations in tumour suppressor genes. Thereby mutant larval brain tissue is implanted into the abdomen of an adult host fly. Wild-type tissue does not overgrowth while benign hyperplasia only grows slowly and does not invade other tissues. In contrast, malignant neoplasm reveals overproliferation and the ability to invade distant organs, which ultimately results in lethality of the host. Malignant neoplasms frequently become immortal and can be expanded up to several months or years through several rounds of implantations into wild-type host flies [32].

Prospero (Pros) is a homeodomain-containing transcription factor that is upon neuroblast division asymmetrically segregated to the GMC by binding the adaptor molecule Mira [34, 35]. In the GMC, Pros is released to the nucleus and acts as a binary switch between self-renewal and differentiation. It then represses genes important for neuroblast self-renewal and activates genes that are involved in cell cycle exit and neuronal differentiation.

Prox1, the vertebrate homologue of Pros, also regulates cell cycle exit in mouse embryonic retinal progenitors and thus Pros/Prox1 function in neural progenitor cells might be conserved from insects to mammals [10, 36, 37].

Brain tumour (Brat) is an NHL domain-containing protein and was first identified by Elizabeth Gateff in a screen for tumour suppressor genes in *Drosophila* [1]. Brat also binds to the adaptor molecule Mira and is thereby inherited by the GMC. Clonal analysis suggests that the Type II neuroblast lineages are particularly susceptible for the loss of Brat function. In this case, the source of the tumour mass derives from intermediate progenitor cells (INP), which fail to mature and overproliferate [22, 37, 38]. Brat can act as a translational repressor and might regulate growth through ribosomal RNA synthesis. Brat was also shown to directly bind and regulate the degradation of specific mRNAs encoding for neuroblast-specific transcription factors Deadpan (Dpn) and Zelda (Zld) [39].

Numb was one of the first asymmetrically partitioned cell fate determinants discovered in *Drosophila* sensory organ precursor and in neuroblasts [40, 41]. It binds to its adaptor molecule Partner of Numb (Pon) and thereby is also inherited by the basal GMC. Numb antagonises Notch signalling and therefore it is likely that Numb acts in the GMC to inhibit Notch activity. Similar to what was observed for Brat, mutant neuroblast clones for Numb fail to differentiate and lead to overproliferation [22].

5.4.2 Examples of Oncogenes Affecting Neuroblast Proliferation

Oncogenes might be important for the ability of the neuroblast to self-renew under normal conditions. However, oncogenes can cause tumourous overgrowth when they are aberrantly activated or overexpressed. *Atypical Proteinkinase C (aPKC)* and *Notch* both fulfil the criteria and are exemplary for oncogenes with a function in *Drosophila* neuroblasts.

5.4.2.1 atypical Proteinkinase C (aPKC)

aPKC is part of the Par complex and regulates neuroblast self-renewal and polarity. Apical aPKC regulates neuroblast self-renewal by restricting GMC fate determinants to the basal cortex. In order to illustrate the oncogenic potential of aPKC, a protein fusion with membrane-tethered CAAX was overexpressed in larval brain neuroblasts. Ectopic localisation of aPKC around the entire cell cortex is tumourigenic and leads to a significant increase in the number of larval neuroblasts [42].

5.4.2.2 Notch (N)

Similarly, neuroblast-specific overexpression of the intracellular domain of Notch leads to massive overgrowth of neuroblasts in larval brains. Dpn might be encoded by one of the Notch target genes that is responsible for the overproliferation phenotype [43–45].

5.4.3   Misregulated Spindle Orientation Can Result in Neuroblast Overproliferation

Mushroom body defective (Mud) has a crucial function for proper spindle orientation. It can bind microtubules and co-localises with Pins at the apical cortex of neuroblasts. *Mud* loss-of-function neuroblasts fail to align their mitotic spindle with the axis of cortical polarity. As a consequence, neuroblasts divide with an orthogonal spindle axis and cell fate determinants segregate symmetrically to both sibling cells. Observations with these aberrant symmetric divisions led to the conclusion that the inheritance of apical proteins correlates 100% with neuroblast identity, whereas the inheritance of basal proteins is insufficient to induce neuronal differentiation [46].

5.5   Brain Tumours Derived from Neuroepithelial Cells

Whereas the mutations in tumour suppressor genes such as *pros*, *brat* and *numb* affect Type I and Type II neuroblast lineages it was discovered that mutations in the other tumour suppressors specifically result in aberrant neuroepithelial proliferation. An initial mutagenesis screen by Elisabeth Gateff identified the gene *lethal (3) malignant brain tumour (l(3)mbt)* as tumour suppressor in larval brains and imaginal discs [47]. L(3)mbt is a conserved transcriptional regulator that binds to cell cycle genes (i.e., E2F, Rb, Cyclin E). Interestingly, in *l(3)mbt* mutant brains tumour formation is specifically initiated by the overproliferation of neuroepithelial cells in the optic lobe. Eventually, the neuroepithelial cells differentiate into an excessive amount of Dpn-positive neuroblasts [48].

5.6   Models for Metastasis in *Drosophila*

Drosophila has become a model to study the metastatic behaviour of tumourigenic cells [49]. Allografts are a valuable approach to maintain and expand the tumour mass for biochemical analysis and metastasis modelling. Another approach is to induce MARCM or Flip-out clones to screen for mutations in tumour suppressor genes or to ectopically express oncogenes. Pagliarini and Xu made the striking observation that loss-of-function mutations in genes encoding cell polarity factors such as Scribbles (Scrib), Discs large (Dlg) or Lethal giant larvae (Lgl) lead to disruption in epithelial polarity and to modest overproliferation phenotypes in eye imaginal disc tissue and optic lobes (Fig. 5.9) [50]. Most overproliferating tissue will eventually differentiate or will be removed by induced cell death (apoptosis). On the other hand, clonal overexpression of an activated form of Ras (Ras^{V12}) leads also to a mild overproliferation phenotype. However, it is the combination of epithelial polarity disruption (i.e., *scrib loss-of-funtion*) together with the activation of oncogenic Ras that results in highly aggressive and invasive tumours. These tumour cells of epithelial origin reveal metastatic properties such as degradation of the basal membrane,

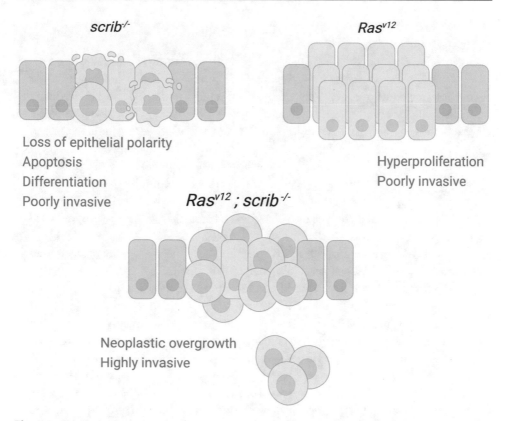

Fig. 5.9 Fly models for *in situ* tumour invasion and metastasis. Scribble (Scrib) deficient clones of cells are removed by dell death, Ras^{V12} expressing cells show hyperplastic outgrowths. When combined, *Ras^{V12}; scrib−/−* cells display severe overgrowth and invasive behaviour. Redrawn after [49], created with BioRender.com

the downregulation of the cell adhesion molecule E-cadherin, migration, invasion and secondary tumour formation [50].

Interestingly, human tumour cells typically lose cell polarity markers and epithelial structure during epithelial to mesenchymal transition. The findings from *Drosophila* relate well to observations made in human metastasis. It reveals the potential that similar screens will uncover genes and molecular mechanisms that are relevant to malignancy in human tumours.

Questions

1. Explain the normal cellular mechanisms that are used to control the size of neural precursor lineages? Give examples from the *Drosophila* larval brain.
2. You have identified a mutation in a tumour suppressor gene that acts in the *Drosophila* brain. How would you proceed to learn more about this gene?

Take-Home Message

Stem cells have the ability to continuously proliferate through self-renewing divisions and at the same time to generate differentiating progeny cell types. While cycling stem cells avoid cell cycle exit and differentiation, they also have to avoid uncontrolled proliferation and tumour formation. More and more evidence suggest that stem cells are likely candidates to initiate tumour growth. Alternatively, any somatic cell might be reverted back to a stem cell-like state thereby becoming immortal, mitotically active and invasive. The cancer stem cell model suggests that only a fraction of cells within the tumour mass remain undifferentiated and are the driving 'malignant' force. Hence, tumour growth might be regarded as an abnormal developing tissue that consists of faulty stem cells together with more differentiated cell types.

Drosophila neuroepithelial cells and neuroblasts are very well characterised neural stem cells and can serve to understand the molecular mechanisms underlying self-renewal, proliferation and differentiation. We have learned that impaired cell fate determination during self-renewing symmetric and asymmetric stem cell division can cause tumourous overgrowth and lead to malignant neoplasms that are invasive and form secondary tumours. Hallmarks of human cancers are recapitulated in the *Drosophila* model system. Many of the tumour suppressor genes and oncogenes that encode cell fate determinants are evolutionary conserved in humans. Thus, *Drosophila* neural stem cells can serve as model to gain molecular, cellular and conceptual insights into how normal stem cells divert from their normal developmental path and adopt tumourigenic properties.

References

1. Gateff E. Malignant neoplasms of genetic origin in *Drosophila* melanogaster. Science. 1978;200(4349):1448–59.
2. Miles WO, Dyson NJ, Walker JA. Modeling tumor invasion and metastasis in *Drosophila*. Dis Model Mech. 2011;4(6):753–61.
3. Knoblich JA. Asymmetric cell division: recent developments and their implications for tumour biology. Nat Rev Mol Cell Biol. 2010;11(12):849–60.
4. Hanahan D, Weinberg RA. Hallmarks of cancer: the next generation. Cell. 2011;144(5):646–74.
5. Egger B, Chell JM, Brand AH. Insights into neural stem cell biology from flies. Philos Trans R Soc Lond Ser B Biol Sci. 2008;363(1489):39–56.
6. Zhang R, Engler A, Taylor V. Notch: an interactive player in neurogenesis and disease. Cell Tissue Res. 2018;371(1):73–89.
7. Kaltschmidt JA, Brand AH. Asymmetric cell division: microtubule dynamics and spindle asymmetry. J Cell Sci. 2002;115(Pt 11):2257–64.
8. Rebollo E, Roldan M, Gonzalez C. Spindle alignment is achieved without rotation after the first cell cycle in *Drosophila* embryonic neuroblasts. Development. 2009;136(20):3393–7.

9. Doe CQ. Neural stem cells: balancing self-renewal with differentiation. Development. 2008;135(9):1575–87.

10. Choksi SP, Southall TD, Bossing T, Edoff K, de Wit E, Fischer BE, et al. Prospero acts as a binary switch between self-renewal and differentiation in *Drosophila* neural stem cells. Dev Cell. 2006;11(6):775–89.

11. Siegrist SE, Doe CQ. Extrinsic cues orient the cell division axis in *Drosophila* embryonic neuroblasts. Development. 2006;133(3):529–36.

12. Yoshiura S, Ohta N, Matsuzaki F. Tre1 GPCR signaling orients stem cell divisions in the *Drosophila* central nervous system. Dev Cell. 2012;22(1):79–91.

13. Maurange C, Gould AP. Brainy but not too brainy: starting and stopping neuroblast divisions in *Drosophila*. Trends Neurosci. 2005;28(1):30–6.

14. Sousa-Nunes R, Cheng LY, Gould AP. Regulating neural proliferation in the *Drosophila* CNS. Curr Opin Neurobiol. 2010;20(1):50–7.

15. Maurange C. Temporal specification of neural stem cells: insights from *Drosophila* neuroblasts. Curr Top Dev Biol. 2012;98:199–228.

16. Chell JM, Brand AH. Nutrition-responsive glia control exit of neural stem cells from quiescence. Cell. 2010;143(7):1161–73.

17. Sousa-Nunes R, Yee LL, Gould AP. Fat cells reactivate quiescent neuroblasts via TOR and glial insulin relays in *Drosophila*. Nature. 2011;471(7339):508–12.

18. Bello BC, Hirth F, Gould AP. A pulse of the *Drosophila* Hox protein Abdominal-A schedules the end of neural proliferation via neuroblast apoptosis. Neuron. 2003;37(2):209–19.

19. Homem CC, Steinmann V, Burkard TR, Jais A, Esterbauer H, Knoblich JA. Ecdysone and mediator change energy metabolism to terminate proliferation in *Drosophila* neural stem cells. Cell. 2014;158(4):874–88.

20. Maurange C, Cheng L, Gould AP. Temporal transcription factors and their targets schedule the end of neural proliferation in *Drosophila*. Cell. 2008;133(5):891–902.

21. Boone JQ, Doe CQ. Identification of *Drosophila* type II neuroblast lineages containing transit amplifying ganglion mother cells. Dev Neurobiol. 2008;68(9):1185–95.

22. Bowman SK, Rolland V, Betschinger J, Kinsey KA, Emery G, Knoblich JA. The tumor suppressors Brat and Numb regulate transit-amplifying neuroblast lineages in *Drosophila*. Dev Cell. 2008;14(4):535–46.

23. Bello BC, Izergina N, Caussinus E, Reichert H. Amplification of neural stem cell proliferation by intermediate progenitor cells in *Drosophila* brain development. Neural Develop. 2008;3:5.

24. Siegrist SE. Termination of *Drosophila* mushroom body neurogenesis via autophagy and apoptosis. Autophagy. 2019;15(8):1481–2.

25. Green P, Hartenstein AY, Hartenstein V. The embryonic development of the *Drosophila* visual system. Cell Tissue Res. 1993;273(3):583–98.

26. Hofbauer A, Campos-Ortega JA. Proliferation pattern and early differentiation of the optic lobes in *Drosophila melanogaster*. Rouxs Arch Dev Biol. 1990;198:264–74.

27. Egger B, Boone JQ, Stevens NR, Brand AH, Doe CQ. Regulation of spindle orientation and neural stem cell fate in the *Drosophila* optic lobe. Neural Develop. 2007;2:1.

28. Yasugi T, Umetsu D, Murakami S, Sato M, Tabata T. *Drosophila* optic lobe neuroblasts triggered by a wave of proneural gene expression that is negatively regulated by JAK/STAT. Development. 2008;135(8):1471–80.

29. Lanet E, Gould AP, Maurange C. Protection of neuronal diversity at the expense of neuronal numbers during nutrient restriction in the *Drosophila* visual system. Cell Rep. 2013;3(3):587–94.

30. Wu JS, Luo L. A protocol for mosaic analysis with a repressible cell marker (MARCM) in *Drosophila*. Nat Protoc. 2006;1(6):2583–9.

31. Lee T, Luo L. Mosaic analysis with a repressible cell marker for studies of gene function in neuronal morphogenesis. Neuron. 1999;22(3):451–61.
32. Gonzalez C. Spindle orientation, asymmetric division and tumour suppression in *Drosophila* stem cells. Nat Rev. 2007;8(6):462–72.
33. Janssens DH, Lee CY. It takes two to tango, a dance between the cells of origin and cancer stem cells in the *Drosophila* larval brain. Semin Cell Dev Biol. 2014;28:63–9.
34. Spana EP, Doe CQ. The Prospero transcription factor is asymmetrically localized to the cell cortex during neuroblast mitosis in *Drosophila*. Development. 1995;121(10):3187–95.
35. Matsuzaki F, Ohshiro T, Ikeshima-Kataoka H, Izumi H. Miranda localizes Staufen and Prospero asymmetrically in mitotic neuroblasts and epithelial cells in early *Drosophila* embryogenesis. Development. 1998;125(20):4089–98.
36. Dyer MA, Livesey FJ, Cepko CL, Oliver G. Prox1 function controls progenitor cell proliferation and horizontal cell genesis in the mammalian retina. Nat Genet. 2003;34(1):53–8.
37. Betschinger J, Mechtler K, Knoblich JA. Asymmetric segregation of the tumor suppressor brat regulates self-renewal in *Drosophila* neural stem cells. Cell. 2006;124(6):1241–53.
38. Bello B, Reichert H, Hirth F. The brain tumor gene negatively regulates neural progenitor cell proliferation in the larval central brain of *Drosophila*. Development. 2006;133(14):2639–48.
39. Reichardt I, Bonnay F, Steinmann V, Loedige I, Burkard TR, Meister G, et al. The tumor suppressor Brat controls neuronal stem cell lineages by inhibiting Deadpan and Zelda. EMBO Rep. 2018;19(1):102–17.
40. Knoblich JA, Jan LY, Jan YN. Asymmetric segregation of Numb and Prospero during cell division. Nature. 1995;377(6550):624–7.
41. Spana EP, Doe CQ. Numb antagonizes Notch signaling to specify sibling neuron cell fates. Neuron. 1996;17(1):21–6.
42. Lee CY, Robinson KJ, Doe CQ. Lgl, Pins and aPKC regulate neuroblast self-renewal versus differentiation. Nature. 2006;439(7076):594–8.
43. Song Y, Lu B. Regulation of cell growth by Notch signaling and its differential requirement in normal vs. tumor-forming stem cells in *Drosophila*. Genes Dev. 2011;25(24):2644–58.
44. San-Juan BP, Baonza A. The bHLH factor deadpan is a direct target of Notch signaling and regulates neuroblast self-renewal in *Drosophila*. Dev Biol. 2011;352(1):70–82.
45. Magadi SS, Voutyraki C, Anagnostopoulos G, Zacharioudaki E, Poutakidou IK, Efraimoglou C, et al. Dissecting Hes-centred transcriptional networks in neural stem cell maintenance and tumorigenesis in *Drosophila*. Development. 2020;147(22)
46. Cabernard C, Doe CQ. Apical/basal spindle orientation is required for neuroblast homeostasis and neuronal differentiation in *Drosophila*. Dev Cell. 2009;17(1):134–41.
47. Wismar J, Loffler T, Habtemichael N, Vef O, Geissen M, Zirwes R, et al. The *Drosophila melanogaster* tumor suppressor gene *lethal(3)malignant* brain tumor encodes a proline-rich protein with a novel zinc finger. Mech Dev. 1995;53(1):141–54.
48. Richter C, Oktaba K, Steinmann J, Muller J, Knoblich JA. The tumour suppressor L(3)mbt inhibits neuroepithelial proliferation and acts on insulator elements. Nat Cell Biol. 2011;13(9):1029–39.
49. Stefanatos RK, Vidal M. Tumor invasion and metastasis in *Drosophila*: a bold past, a bright future. J Genet Genomics. 2011;38(10):431–8.
50. Pagliarini RA, Xu T. A genetic screen in *Drosophila* for metastatic behavior. Science. 2003;302(5648):1227–31.

Eye Development in *Drosophila*: From Photoreceptor Specification to Terminal Differentiation

6

Abhishek Kumar Mishra, Simon G. Sprecher

What You Will Learn in This Chapter

This chapter highlights different steps of *Drosophila* compound eye development starting from early photoreceptor specification in the embryo to the terminal differentiation and determination of photoreceptor subtype specificity. In this chapter, "retinal determination cascade" is briefly described and it also provides details of how crosstalk of retinal determination genes at different levels regulates the eye developmental process. This chapter also highlights "morphogenetic furrow" initiation and progression in the eye-antennal imaginal disc and how that leads to the specification of different photoreceptor subtypes in the compound eye. It also provides details of photoreceptor terminal differentiation and how different ommatidial subtypes are determined and arranged in the compound eye.

6.1 *Drosophila* as a Model Organism to Understand Eye Development

The ability to sense and interpret information from the visual world is processed through the eyes that perceive information from its surroundings and process visual cues to the brain allowing us to react appropriately. The predominant view of the eye development in the twentieth century was polyphyletic origin of eye and it was believed that eyes

A. K. Mishra · S. G. Sprecher (✉)
Department of Biology, University of Fribourg, Fribourg, Switzerland
e-mail: simon.sprecher@unifr.ch

© Springer Nature Switzerland AG 2023
B. Egger (ed.), *Neurogenetics*, Learning Materials in Biosciences,
https://doi.org/10.1007/978-3-031-07793-7_6

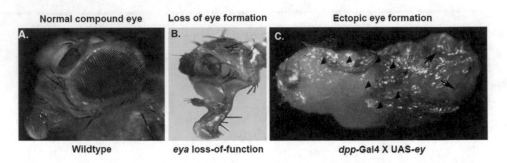

Fig. 6.1 (a) Wildtype *Drosophila* compound eye (b) Loss of compound eye formation in *eya* loss-of-function mutant (c) Ectopic eye formation upon forced expression of eyeless (ey) by *dpp*-Gal4. Black arrow represents compound eye whereas black arrowhead represents ectopic eyes. Modified from [15]

evolved several times independently [1]. However, this view was challenged by Walter Gehring's group from Switzerland in the mid-1990s when they discovered *eyeless* (*ey*) gene in *Drosophila* is homologous to the mouse *Pax6* gene [2]. It was already known that mutation in human *Pax6* gene cause aniridia, a retinal disorder where eye development is severely hampered [3, 4]. Gehring's group further demonstrated that forced expression of *ey* in non-retinal tissues could transform them into an eye fate [5]. Further research in this direction showed that Pax6/*ey* homologs are actually present in all major animal phyla and regulate eye development in all cases, making it as a "master regulator" gene [6–14]. In subsequent years, more genes were added to the list of "master regulator genes of eye development" and a gene regulatory network called "retinal determination network (RDN)" was established. The members of RDN gene family were either absolutely required for eye development or ectopic eye formation upon forced expression or both (Fig. 6.1) [16, 17]. The *Drosophila* compound eye serves as a remarkable model to understand mechanisms of organogenesis, cell proliferation, apoptosis, cell fate specification, pattern formation, and morphogenesis.

6.2   Morphology of the *Drosophila* Compound Eye

The compound eyes in *Drosophila* contain hexagonal array of unit eyes called ommatidia. Each compound eye contains approximately 800 ommatidia organized in a neurocrystalline lattice [18]. Each ommatidium harbors a core of eight light-sensing photoreceptor (PR) neurons surrounded by four non-neuronal cone cells that secrete corneal lens and two primary pigment cells. Out of eight PR neurons, six PRs (R1-R6) are also called outer PRs and they are marked by large rhabdomeres and expression of blue-sensitive Rh1. Outer PRs project their axons into the lamina, a region in the brain that is mainly involved in motion detection and dim light vision [19–21]. R7 and R8 are also called inner PRs, which are marked by centrally located small rhabdomeres, with R7 sitting just above R8 and their

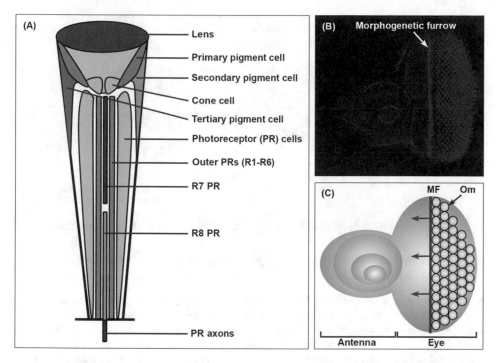

Fig. 6.2 Morphology of the compound eye. (**a**) Each eye consists of approx. 800 ommatidia and each ommatidium (one is represented here) consists of eight PRs. R1–R6 (colored in gray) are considered as outer PRs whereas R7 (colored in magenta) and R8 (colored in light blue) are considered as inner PRs. The PRs are surrounded by pigment cells (primary, secondary, and tertiary) which acts as light insulators. Lens and cone cells act to establish the correct optical path for light detection by PRs. (**b**) Representative image of an eye-antennal imaginal disc of a third instar larvae where "white arrow" represents morphogenetic furrow (MF) and "blue cells" represent differentiated PRs. (**c**) Schematic representation of a third instar eye-antennal imaginal disc where "red border" represents MF and "red arrows" represent direction of the MF from posterior to anterior side of the disc. Anterior part of the disc transform and become the antenna whereas posterior part transforms and become the eye. Om, ommatidium. Modified from [22]

PRs express a combination of UV-(Rh3, Rh4) and blue-(Rh5) or green-(Rh6) sensitive *Rhodopsins*. Inner PRs project their axons into the medulla region of the brain that are responsible for color vision (Fig. 6.2) [23, 24]. The ommatidial clusters are separated by secondary and tertiary pigment cells and mechanosensory bristles that limit light scattering [18, 25].

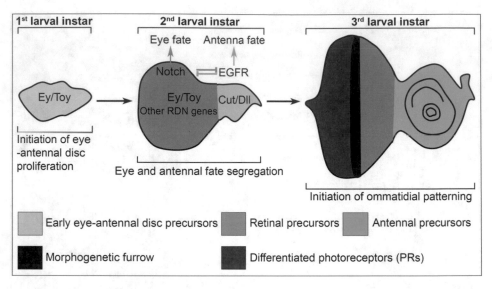

Fig. 6.3 Schematic representation of eye and antennal fate segregation in the eye-antennal imaginal disc during different larval stages. Initiation of the eye-antennal imaginal disc proliferation occurs at the first larval instar whereas eye and antennal fate segregation occurs at the end of second larval instar. Ey and Toy is initially expressed in the entire eye-antennal imaginal disc at the first larval instar but gets restricted to the presumptive eye part of the second larval instar. Similarly, cut and Dorsal (Dl) are expressed in the presumptive antennal part of the disc. Antagonistic role of EGFR and Notch signalling provides eye and antennal fate and after eye and antennal fate are segregated, other RDN genes also start to get expressed at the late second larval instar. At the third larval instar, ommatidial patterning occurs as a result of PR differentiation

6.3 Determination of Eye and Antennal Fate in the Eye-Antennal Imaginal Disc

Upon hatching into a larva, the presumptive cells of the eye-antennal imaginal disc, which were set aside during embryogenesis undergo rapid proliferation and get self-organized into a monolayer epithelium. At the larval first instar, eye-antennal disc epithelium only consists of small cluster of cells with no regional identity (Fig. 6.3). During first and second larval instars, the monolayer epithelium divides rapidly and generates requisite number of cells required to form an ommatidial assembly. Next, they induce the expression of tissue-specific regulatory genes within distinct regions of eye and antennal field to subdivide the eye-antennal disc primordia (Fig. 6.3). Eye field in the disc is marked by expression of retinal determining genes that interplay with signaling pathways to induce growth and development of the eye disc [26–31]. The eye region of the eye-antennal imaginal disc grows and form the eye, head cuticle, and ocelli whereas the antennal region grows to form the antenna and head cuticle (Fig. 6.3) [32]. Initiation of RDN gene expression to segregate eye and antennal fate in the eye-antennal disc begin at late larval second instar.

The core member of RDN *eyeless (ey)* and *twin-of-eyeless (toy)* initially gets expressed in the entire eye-antennal imaginal disc primordia at larval first instar. While eye fate is determined by *ey* that controls other RDN genes, antennal fate is similarly determined by the expression of homeodomain transcription factors Homothorax (Hth), Distal-less (Dll) and Cut. During late larval second instar, *ey/toy* expression is restricted to the two-third of the posterior disc whereas *cut* expression is dominated in the anterior third of the eye-antennal disc. *ey/toy* expressing domain marks the region of future eye whereas *cut* domain signifies antennal region in the eye-antennal disc. EGFR and Notch signalling pathways play important roles in providing regional identity (eye and antenna) in the eye-antennal imaginal disc. EGFR signalling represses *ey* expression in the antennal part, whereas Notch signalling antagonises EGFR pathway in the eye part in order for eye development to proceed (Fig. 6.3) [33].

6.4   The Retinal Determination Cascade

According to the current state of research, the *Drosophila* eye is determined by the activity of approximately 14 genes and 5 signalling pathways. A vast majority of these genes are transcription factors, and they make up the "traditional" eye specification network called retinal determination network (RDN) to regulate eye development process at multiple stages. Most of the RDN genes has a vertebrate counterpart (Fig. 6.4) and they are known to be involved in several clinical retinal disorders [35]. In flies, it includes *eyeless (ey)*, *twin of eyeless (toy)*, *eyegone (eyg)*, *twin of eyegone (toe)*, *sine oculis (so)*, *optix*, *teashirt (tsh)*, *tiptop (tio)*, *distal antenna (dan)*, *distal antenna related (danr)*, *dachshund (dac)*,

Drosophila genes	Vertebrate homolog(s)	Functional domain(s)
eyeless (ey)	Pax6	Paired/homeodomain
twin of eyeless (toy)	Pax6	Paired/homeodomain
sine oculis (so)	Six1/2	Homeodomain
Optix	Six3/6	Homeodomain
eyes absent (eya)	Eya1-4	Tyrosine phosphatase
dachshund (dac)	Dach1-2	Winged helix-turn-helix
eyegone (eyg)	Pax6(5a)	Paired/homeodomain
twin of eyegone (toe)	Pax6(5a)	Paired/homeodomain
homothorax (hth)	Meis1	TALE homeodomain
teashirt (tsh)	TshZ1-4	Zinc finger
tiptop (tio)	TshZ1-4	Zinc finger
distal antenna (dan)	---	Pipsqueak
distal antenna related (danr)	---	Pipsqueak
nemo (nmo)	Nlk	Serine/threonine kinase

Fig. 6.4 A list of retinal determination network (RDN) genes in *Drosophila*, its vertebrate homologs and their known functional domains. Modified from [34]

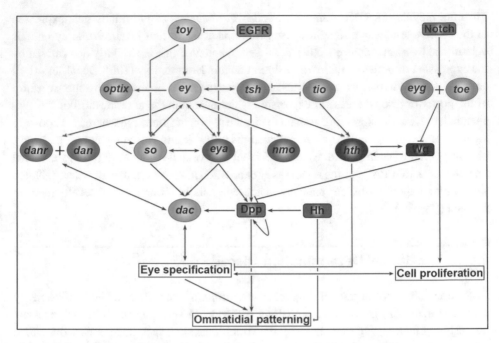

Fig. 6.5 A schematic of retinal determination cascade that leads to eye specification and ommatidial patterning in *Drosophila*. Blue arrows marks activation whereas red lines marks repression. Modified from [34]

homothorax (*hth*), *eyes absent* (*eya*), and *nemo* (*nmo*) [2, 26–30, 36–44]. Eya belongs to a family of protein tyrosine phosphatase and acts as a transcriptional activator whereas Nemo belongs to the protein kinase family (Fig. 6.4). Genes in the RDN family control multiple tasks such as cell proliferation, eye specification by initiation and migration of the morphogenetic furrow, ommatidial patterning, and apoptosis by reinforcing mutual negative interactions and autoregulatory feedback mechanisms (Fig. 6.5) [45, 46]. They are also integrated with signalling pathways at multiple points during eye development and these signalling pathways include the Notch, EGFR, Hedgehog, TGFß and Wnt pathways (Fig. 6.5). While some of these pathways promote eye development, others appear to switch between promoting and inhibiting eye development process depending on their spatiotemporal appearance [33, 47–51].

6.5 Morphogenetic Furrow

The eye develops from a monolayer epithelial tissue known as eye-antennal imaginal disc that invaginates in the embryo, grows and proliferates in the larval stages and retinal differentiation starts in the third instar larva. While the anterior part of the disc grows

Fig. 6.6 The morphogenetic furrow (MF) and arrangement of cells at its anterior and posterior side. Anterior side of the furrow contains differentiated PRs. Ommatidial assembly can be seen at the posterior side of the furrow. Adherens junctions of the epithelium is marked in red whereas differentiating PRs are shown in blue. Cell membrane of the differentiating PRs are marked in green. Modified from [22]

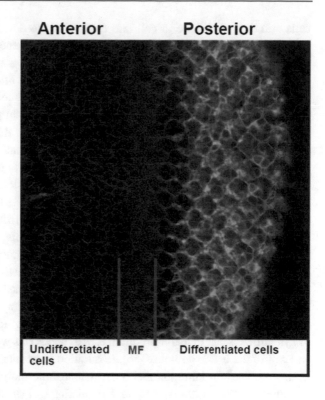

Anterior   Posterior

Undifferetiated   MF   Differentiated cells
cells

and becomes the antenna, the posterior part grows and forms the eye and additional head cuticles. Epithelial-to-neural transition occurs as a wave in the third instar larvae resulting in the retinal differentiation from posterior to anterior end of the disc and that appears as a physical indentation known as "morphogenetic furrow" (Fig. 6.6). Morphogenetic furrow moves from posterior to anterior side of the eye disc. Anterior to this moving furrow, undifferentiated cells are arranged in an unpatterned manner and on the posterior side, patterning occurs that results in the formation of highly organized ommatidial clusters (Fig. 6.6). Once the entire eye field is established, terminal differentiation occurs and specific PR cell types are formed [25].

6.6   Initiation of the Morphogenetic Furrow

When the larva enters into the third and final instar stage, a wave of differentiation called morphogenetic furrow is initiated at the posterior region of the disc and it moves towards the anterior side [18]. This movement of the furrow transforms unpatterned and undifferentiated cells into highly organized ommatidial clusters and triggers RDN gene expression to undifferentiated and possibly determined cells [2, 18, 28, 39, 44, 52–56]. Few genes of the RDN such as *so, eya, dac*, and *nmo* are also expressed in differentiated cells and they contribute in cell fate choices in the developing ommatidia [26, 29, 30, 37, 38]. Based on the expression of different RDN genes, the eye domain of the eye-antennal imaginal disc is subdivided into 6 expression zones (Fig. 6.8) [34, 46, 57]. The most anterior zone (Zone A) expresses the RDN gene *hth* and its co-factor *exd* [42, 58, 59] and this zone does not produce any retinal tissues but it gives rise to the head cuticle [32]. In addition to *hth* and *exd*, *cut* is also expressed in part of this zone and it acts as an additional signal to block retinal development [60]. Adjacent to Zone A lies Zone B and cells in Zone B are involved in adopting an eye fate. Cells in this zone are highly proliferative in nature and express RDN genes such as *ey/toy*, *eyg/toe*, *tsh/tio*, *dan/danr* and *optix* [2, 28, 36, 39, 43, 44, 52, 53]. Zone C (also called pre-proneural zone) are closest to the morphogenetic furrow and while cells in this zone still proliferate, they are on the edge to get incorporated into the ommatidial clusters [61]. Zone C additionally express *so, eya* and *dac* and they help in the transition of cells from being undifferentiated to adopting retinal cell fate by initiating and progressing morphogenetic furrow [29, 62, 63]. Zone D lies within morphogenetic furrow itself where cells are arrested in G1 phase of the cell cycle and the G1 arrest is established and maintained by high levels of TGFß ligand Decapentaplegic (Dpp) [64]. As the morphogenetic furrow moves, subset of cells exits the cell cycle and begins to specify in the following order: R8-R2/R5-R3/R4 and one last round of synchronous cell division produces R1/R6 and R7 as well as cone cells, pigment cells and cell of the bristle complex [56]. In addition to maintaining G1 arrest, Dpp signalling is also required for initiation and progression of the morphogenetic furrow [65] and one possible hypothesis is that it is through regulating *so, eya*, and *dac* expression just ahead of the furrow. Unfortunately, it is shown that Dpp regulates *so, eya*, and *dac* expression only at the margin of the disc during morphogenetic furrow initiation [66, 67]. The area behind the furrow is subdivided into zone E and zone F. Zone E can be molecularly identified by the expression of *so, eya, dac*, and *nmo* whereas Zone F is the posterior-most zone of the eye field and can be identified by the expression of *so, eya*, and *nmo* (Fig. 6.7).

Signalling pathways such as the Hedhehog (Hh), Wingless (Wg), Decapentaplegic (Dpp), Notch and EGFR also provide a basis for the initiation and propagation of the morphogenetic furrow [66, 68]. The birth of the furrow occurs at the posterior side of the eye disc and its thickness increases dorsoventrally during the furrow movement. Morphogenetic furrow formation occurs repeatedly at the rim of the disc and it is called furrow reincarnation [68]. Hh and Dpp signalling acts to promote furrow initiation and reincarnation whereas Wg signal acts an inhibitor [51]. Hh is expressed at the posterior margin before initiation of the furrow and lack of Hh activity results in the complete inhibition of furrow formation whereas its forced expression ahead of the furrow leads to ectopic furrow formation [51, 69]. Dpp is expressed along the posterior and lateral margin of the eye disc and ectopic Dpp activity results in precocious furrow formation whereas loss of Dpp signal leads to the loss of morphogenetic furrow (Fig. 6.8). Additionally, Dpp activates *eya, so*, and *dac* expression and thereby regulates the expression of RDN genes [66]. Wg is expressed at the lateral margins just anterior to the morphogenetic furrow and its ectopic expression controls morphogenetic furrow progression whereas loss of Wg

Fig. 6.7 A Schematic representation of RDN gene's expression pattern in the eye part of eye-antennal imaginal disc during third instar larvae. Posterior region of the disc acquires eye fate whereas anterior region grows and form the antenna. Based on expression pattern of RDN genes, the developing eye disc can be subdivided into six expressing zones (Zone A-F; represented in different colors). The morphogenetic furrow (MF) zone is colored in gray and black arrows within this zone show MF movement from the posterior to anterior side of the disc. Modified from [34]

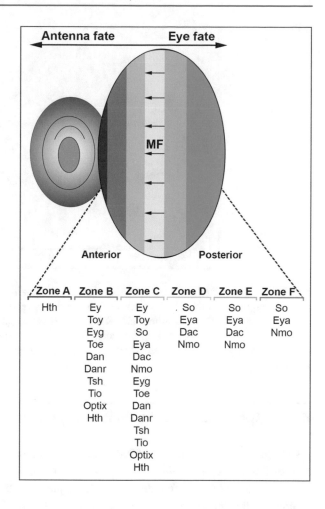

activity leads to ectopic furrow formation. Wg signal blocks the expression of *so, eya,* and *dac* indirectly by activating *hth* expression in the eye-antennal imaginal disc [31, 58]. Antagonistic role of Notch and EGFR signalling promotes eye fate and furrow initiation [68]. Notch signalling activates and maintains *ey* expression whereas EGFR signalling restricts *ey* expression in the developing eye (Fig. 6.8) [33, 50]. These signalling pathways not only promote furrow initiation but are also required for furrow movement across the eye epithelium. Hh is expressed in the differentiating photoreceptors behind the furrow and it activates Dpp at the anterior edge to propagate the progression of "pre-proneural" state to the "proneural" state [67, 70]. Pre-proneural cells express *hairy* whereas proneural cells are marked by *atonal* (*ato*) expression. Hh, Dpp, and Notch signalling are required for repression of *hairy* expression in order to promote *ato* expression.

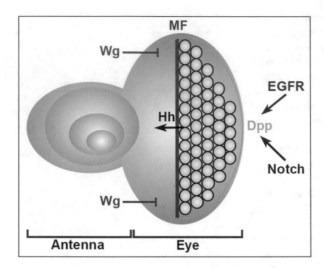

Fig. 6.8 A representative model to show roles of signalling pathways in the initiation and propagation of MF at the eye disc. MF initiation and propagation requires an interplay of Wg, Hh, EGFR, Notch, and Dpp signalling. EGFR and Notch participate in initiating the furrow whereas Wg acts as a negative signal. Dpp promotes furrow reincarnation at the lateral margins whereas reiterative propagation of the furrow depends on an interplay between Hh and Dpp signalling pathway. Modified from [22]

6.7   Specification of R8 Photoreceptor Neuron

As the progressive movement of morphogenetic furrow occurs from posterior to anterior side of the disc, a new column forms every 2 hours. The first PR neuron to be specified from the furrow is R8 founder cells. Subsequently, R8 recruits other PR precursors. Therefore, the initial specification of R8 is very critical since no other PR precursors will develop if R8 is absent [71, 72]. The proneural gene *ato* is required to specify R8 founder cells. Ato expression is highly dynamic in the eye-antennal imaginal disc. Initially, Ato expression appears as a wide stripe just anterior to the furrow and progressively it becomes more and more restricted until it is expressed only in a single cell per cluster [71]. The broad band of Ato initially splits into a cluster containing about 10 *ato* expressing cells and from there two to three cells move apically and form R8 equivalence group. All cells in the equivalence group are equipotent to give rise to R8 precursors. However, only one cell at the end maintains *ato* expression and finally specified as R8 precursor (Fig. 6.9). Therefore, a tight control of *ato* activity is required for the specification of single R8 precursor per cluster and indeed *ato* activity is tightly controlled by regulatory elements at its 3′ and 5′ enhancers. Initial broad *ato* expression is controlled by regulatory elements at its 3′ enhancer whereas later *ato* expression is controlled by distinct 5′ element. Signalling pathways such as Notch, Hh, and Dpp act to ensure that only once cell per

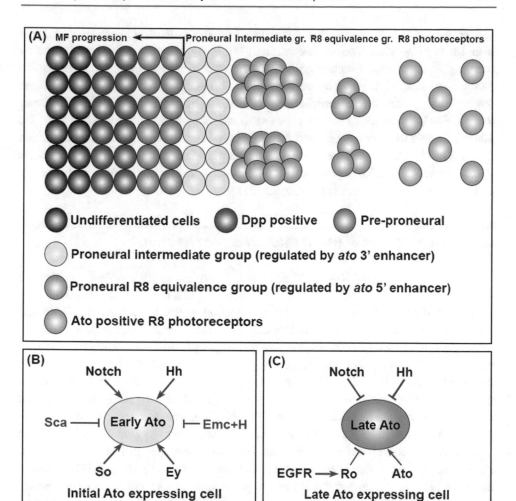

Fig. 6.9 Model for the specification of R8 PRs during ommatidial development. (**a**) Specification of R8 start with MF movement where undifferentiated cells (colored in gray) first start to express Dpp (cells colored in magenta) and then acquire "pre-proneural state" (cells colored in purple). Maturation of the "pre-proneural" to "proneural" state occurs and proneural cells start to express *atonal (ato)* (colored in yellow). Initial *ato* expression gets restricted to intermediate group of cells (8–10 cells) and they later form R8 equivalence group containing 2–3 cells (colored in pink). Initial *ato* expression is regulated by its 3′ enhancer region whereas late *ato* activity is regulated by its 5′ enhancer. (**b**) Early *ato* activity is promoted by Notch and Hh signalling as well as RDN genes such as *so* and *ey* and it is repressed by fibronectin-like protein Scabrous (Sca) and the transcriptional repressors *Hairy (H)* and *Extramacrochaetae (Emc)*. (**c**) Late *ato* expression is maintained by *ato* auto-regulation and it is repressed by Notch and Hh signalling and the transcription factor *Rough (Ro)*. Modified from [22]

cluster is specified. Ato expression is initially promoted by Hh and Notch signalling and later Hh negatively regulates *ato* expression [73]. Early Ato expression is repressed by transcriptional repressors such as *hairy* and *extramachrochaetae* (emc) and promoted by Notch signalling. Notch initially promotes *ato* expression by repressing *hairy* and *emc* whereas during later stages it represses *ato* activity. The homeodomain transcription factor Rough (Ro) is expressed in non-R8 cells due to lateral inhibition mediated by Notch. Ro inhibits R8 fate by repressing *ato* and therefore in *ro* mutants, non-R8 cells adopt R8 fate (Fig. 6.9) [74]. Conversely, the Zn-finger transcription factor Senseless (Sens) promotes R8 fate by repressing Ro expression in the R8 cells. In *sens* mutants, R8 fails to specify and adopt an R2/R5-like fate expressing Ro [71].

6.8   Recruitment of R1–R7 Photoreceptor Neurons

Subsequent recruitment of R1–R7 PR neurons is controlled by R8 and their specification occurs in the following order: first R2 and R5, then R3 and R4, then R1 and R6 and finally R7 (Fig. 6.10) [18, 55]. After the initial specification of R8, recruitment of R2/R5 and R3/R4 occurs rapidly and they form five-cell pre-cluster. Further recruitment of R1/R6, R7, and other accessory cells require post-furrow proliferation also known as "second mitotic wave." A number of signalling events and transcription factor actions are required for proper specification of all PRs. First of all, specification of PRs except R8 is under the influence of EGFR signalling pathway [75, 76]. The EGFR ligand Spitz is produced by R8 and allows the recruitment of other PRs in a stepwise manner [77, 78]. The first pair specified after R8 is R2 and R5, which require *rough* activity (Fig. 6.10) [79]. *rough* is expressed in R2/R5 and R3/R4 pairs and R2/R5 is mis-specified in *rough* mutants. Further, the misexpression of an orphan nuclear receptor Seven-up (Svp) in R2/R5 blocks their specification suggesting that Ro acts to promote R2/R5 specification by repressing Svp [80]. Also, R3/R4 specification requires inputs from R2/R5 since in *rough* mutants, recruitment of R3/R4 is inappropriate in the developing ommatidium [81, 82].

Svp is required for the specification of R3/R4 and R1/R6 since in *svp* mutants, these PRs are transformed into an R7-like state [83]. R3/R4 specification also requires *spalt* gene complex which encodes two Zn-finger transcription factors *spalt major* (*salm*) and *spalt related* (*salr*). Spalt in essential for the activation of Svp expression in R3/R4 [84].

Specification of R1/R6 requires the activity of a transcription factor Lozenge (Lz) which promotes R1/R6 fate by activating the expression of a homeodomain transcription factor BarH1 [85, 86]. The last PR specified within an ommatidium is R7 where *lozenge* activity is required for repressing *svp* expression [85]. R7 specification is also controlled by a receptor tyrosine kinase Sevenless (Sev), which in addition to R7, also expressed in R3/R4, cone cell precursors and R1/R6 [87]. The ligand of Sev, also known as Bride of sevenless (Boss) is specifically expressed in R8. Although Sev is also expressed in R3/R4 and R1/R6 pairs, forced expression of Sev in cone cell precursors transform them as only R7, most likely because in R3/R4 and R1/R6 pairs, Svp is expressed [88]. Loss of *svp* from R3/R4

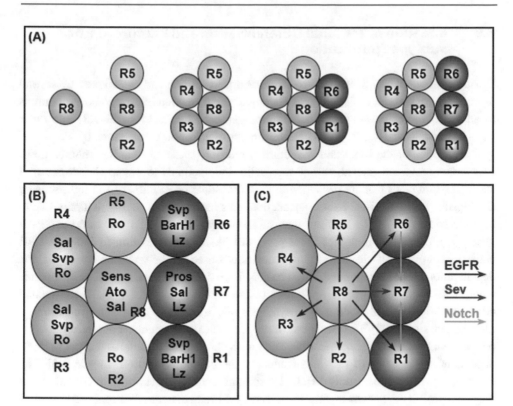

Fig. 6.10 Recruitment of R1 to R7 PRs into the ommatidium. (**a**) Recruitment of R1–R7 from R8 (in blue) strictly follows a stereotypical order: first R2 and R5 (in green), then R3 and R4 (in pink), then R1 and R6 (in purple) and finally R7 (in magenta). (**b**) Specification of different PRs requires distinct transcriptional factor code. R8 expresses Senseless (Sens), Atonal (Ato) and Spalt (Sal) whereas R2/R5 pair expresses Rough (Ro), R3/R4 pair expresses Sal, Seven-up (Svp) and Ro, R1/R6 pair expresses Svp, BarH1 and Lozenge (Lz) and R7 expresses Prospero (Pros), Sal and Lz. (**c**) Recruitment of R1–R6 requires EGFR signalling from R8 PRs whereas R7 specification requires Notch signalling from R1 and R6 and Sevenless (Sev) signalling from R8 PRs. Modified from [22]

and R1/R6 PRs transform them as R7 whereas high levels of *svp* in R7 cells transform them as outer PRs [80, 89, 90]. In addition to Sev signalling, R7 specification also requires Notch activity. Notch signals are provided to R7 by its neighboring cells R1 and R6 and if this signal is lost in R7, they transform and become R1/R6-like cells [91, 92]. Additionally, R7 specification is also controlled by another homeodomain transcription factor *prospero* (*pros*), a gene which is highly expressed in R7 precursors (Fig. 6.10) [93, 94].

6.9   Initiation of Terminal Differentiation and Photoreceptor Subtype Specification

During metamorphosis, PRs undergo terminal differentiation and are marked by several morphological and molecular changes leading to the formation of functional PR neurons in adults. Some of these events include arrangement of eight PRs (R1–R8) into outer (R1–R6) and inner (R7, R8) PRs, establishment of synaptic connections by PR axons into optic lobe of the brain and rhabdomere morphogenesis. The apical membrane of PRs becomes folded into numerous microvilli and modifies to create rhabdomeres, a stack of membranes that are packed with phototransduction machinery and light-sensing rhodopsins [95]. The visual phototransduction machinery convert energy of a photon into PRs electrical responses which are then taken up by the brain and processed as visual ques. [96]. Disruption in any of these processes blocks the formation of functional PR neurons and eventually leads to PR degeneration. Based on the mutational analysis, PR degeneration can be categorized into 4 major classes: The first class consists of mutants where PRs connectivity to the optic lobe of the brain is hampered [97]. In the second class, mutations affect rhodopsin biosynthesis and its maturation during rhabdomeres morphogenesis [95, 98]. The third class of mutations are components of phototransduction pathway that leads to PR degeneration. The fourth class includes mutants that cause light-dependent degeneration of PRs independent of the phototransduction cascade [96]. The outer and Inner PRs are not only different in size and orientation within an ommatidium but they also project their axons to different areas in the brain, contains different rhodopsins as wells as perform different tasks. Outer PRs in fly are equivalent to vertebrate rods and they are mostly implicated in motion detection and dim light vision. They contain a broad range blue-green light-sensitive photopigment Rhodopsin 1 (Rh1) (Fig. 6.11) and display large rhabdomeres that are able to catch photons with high accuracy [19–21]. They project their axons to the lamina cortex of the optic lobe in the brain. Inner PRs in fly are equivalent to vertebrate rods and are mostly implicated in color discrimination [99, 100]. Inner PRs do not span the whole length of an ommatidium but they are arranged on top of each other (R7 on top of R8) and express a range of rhodopsin photopigments such as UV-sensitive Rh3 and Rh4, blue-sensitive Rh5 and green-sensitive Rh6 (Fig. 6.11). Inner PRs diameter are also considerably small than outer PRs and in terms of their axonal projections, they extend their axons into the medulla cortex of the optic lobe in the brain [100].

Although external appearance of ommatidia in the compound eye looks the same, they can be subdivided into three distinct classes: yellow (y) ommatidia, pale (p) ommatidia and ommatidia of the dorsal rim area (DRA) (Fig. 6.11). Yellow and pale ommatidia covers most of the compound eye and they are randomly distributed throughout the eye in approximately 70/30 ratio (70% yellow and 30% pale). Their absorption spectra indicate that they contain different sets of rhodopsins [101, 102]. Indeed, it was found that yellow ommatidia contains Rh4 in R7 and Rh6 in R8 whereas pale ommatidia contains Rh3 in R7 and Rh5 in R8 (Fig. 6.11). Therefore, based on their absorption spectra it is concluded that yellow ommatidia is involved in discriminating between longer wavelength of light

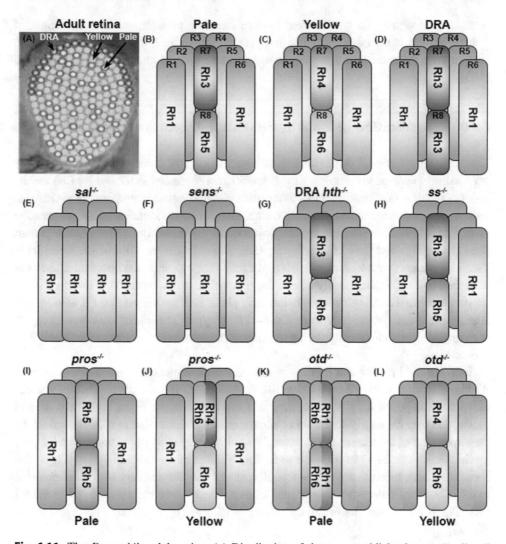

Fig. 6.11 The *Drosophila* adult retina. (a) Distribution of three ommatidial subtypes ("yellow," "pale," and "DRA") in the adult compound eye. (**b–d**) Schematic representation of the yellow, pale and DRA ommatidial subtypes. The outer PRs (R1–R6) of all ommatidial subtypes (represented in brown) express Rh1 and have large rhabdomeres surrounding two inner PRs (R7 and R8; R7 being on top of R8). Inner PRs have smaller rhabdomeres and based on rhodopsin expression they can be subdivided into: (**b**) pale ommatidia (where Rh3 is expressed in R7 (represented in magenta) and Rh5 is expressed in R8 (represented in blue)), (**c**) yellow ommatidia (where Rh4 is expressed in R7 (represented in purple) and Rh6 is expressed in R8 (represented in green)) and (**d**) DRA ommatidia (where Rh3 is expressed in both R7 and R8 (represented in magenta)). (**e**) In $sal^{-/-}$ mutants, inner PRs are transformed, and they become outer PRs expressing Rh1 (**f**) In $sens^{-/-}$ mutants, R8 does not form and all PRs are identical to the outer PRs expressing Rh1 (**g**) In $hth^{-/-}$ mutants, DRA ommatidia gets transformed and become odd-coupled ommatidia expressing Rh3 in R7 and Rh6 in R8. (**h**) In $ss^{-/-}$ mutants, all ommatidia become pale ommatidia expressing only Rh3 in R7 and Rh5 in R8. (**i, j**) In $pros^{-/-}$ mutants, Rh5 and Rh6 are derepressed in R7 but is unaffected in R8. (**k, l**) In $otd^{-/-}$ mutants, Rh6 is derepressed in outer PRs and in some R7 and R8 of the pale ommatidia, together with Rh1. Modified from [22]

whereas pale ommatidia are involved in discriminating between shorter wavelength of light. However, in comparison to pale and yellow ommatidia, DRA ommatidia expresses Rh3 in both R7 and R8 and are implicated in perception of polarized light (Fig. 6.11) [103].

6.10   Specification of Inner Versus Outer Photoreceptor Subtypes

The molecular basis of distinguishing between inner and outer PRs is based on the expression of spalt gene complex. *Salm* is specifically expressed in R7 and R8 PRs and its loss-of-function transform inner PRs to the outer PRs that includes morphological changes, loss of inner PR Rhodopsins (Rh3, Rh4, Rh5, and Rh6) and initiation of Rh1 expression (Fig. 6.11) [104]. However, the axonal projection of the transformed PRs in *salm* mutant does not change and they still terminate their axons in the medulla of the optic lobe. This also means that initial specification of PRs in the eye-antennal imaginal disc is not disturbed and *salm* mutant only affects the terminal differentiation. This finding further subdivides retinal development where the early phase is to get the general PR identity and later terminal differentiation produce distinct PR subtypes. The role of *spalt* gene complex is to provide inner PR identity during terminal differentiation in an otherwise "outer PR" ground state.

Within inner PRs, R7 and R8 display several morphological differences. For example, R7 that is always located on the top of R8, nucleus of R7 is more distal whereas nucleus of R8 is more proximal, R7 project its axons in a deeper layer in the brain than R8 and in addition they express different set of *rhodopsoins*. R7 fate is distinguished from R8 by the expression of the transcription factor Pros, which is specifically expressed in R7 (and not in R8 or outer PRs) in response to EGFR, Notch, and Sev signalling [93]. It is also shown that Pros binds to the enhancer region of *rh5* and *rh6* and repress their expression in R7. Therefore loss-of-function of *pros* leads to de-repression of *rh5* and *rh6* in the R7 PRs (Fig. 6.11) whereas its forced expression in R8 leads to repression of *rh5* and *rh6* in R8 [93].

6.11   Stochastic Determination of Yellow Versus Pale Ommatidia

As mentioned earlier, yellow and pale ommatidia are randomly organized in the compound eye in a 70/30 ratio (Fig. 6.11). The development to form either yellow or pale ommatidia in the compound eye is determined by the transcription factor Spineless (Ss) [105]. Ss is expressed in a subset of R7 PRs in a stochastic manner (expressed in 60–80% of R7) and acts to determine yR7 fate. In *spineless* mutants, all R7 expresses Rh3 and they adopt pR7 fate (Fig. 6.11) whereas its forced expression induces Rh4 expression in all R7 and thereby transforms all R7 into yR7. In spineless mutants, most R8 PRs express Rh5 (Fig. 6.11) whereas spineless forced expression makes R8 to express Rh6. Therefore, Spineless acts at multiple levels during ommatidial specification. First, it promotes Rh4 expression and represses Rh3 in R7 and secondly, in the underlying R8, it promotes Rh6 in a non-autonomous manner. This is also one such example of cell–cell communication to attain cell fate and it shows that Ss expressing yR7 control a signal, which is required by R8 to adopt the yR8 fate [101, 105].

The fate choices between yR8 (expressing Rh6) or pR8 (expressing Rh5) are further controlled by a genetic bistable loop consisting of a growth regulator *warts* (*wts*) and a tumor suppressor *melted* (*melt*). Wts is expressed in yR8 where it promotes Rh6 expression by repressing Rh5 whereas Melt is expressed in pR8 where it promotes Rh5 expression at the expense of Rh6. Both *wts* and *melt* are expressed in a mutually exclusive manner where *wts* represses *melt* and vice versa to regulate fate choices between yR8 and pR8 [106].

It is also known that the homeodomain transcription factor Otd is expressed in PRs and is required for PR morphogenesis in the eye [107]. An eye-specific *otd* allele, *otd*^{uvi} shows significant reduction of Otd expression in the eye and in this allele sensitivity to UV is lost together with severe PR morphological defects [107]. In *otd*^{uvi} flies, Rh4 and Rh6 are normally expressed in yellow ommatidia whereas Rh3 and Rh5 expression is lost from the pale ommatidia [108]. In addition, Rh6 expression is expanded in outer PRs as well as slight derepression of Rh1 in yR7 and yR8 occurs (Fig. 6.11). These observations provide the evidence that: (1) Otd is required for the expression of *rh3* and *rh5* in the inner PRs of pale ommatidia. (2) Otd acts as a repressor of *rh6* in the outer PRs and (3) Ots acts as a repressor of *rh1* in the inner PRs. Further analysis of *rhodopsin* promotors identified potential Otd binding site (K50) in the enhancer region of *rh3*, *rh5*, and *rh6* and it was concluded that Otd indeed binds to the K50 site to regulate *rh3, rh5*, and *rh6* expression. Surprisingly, no K50 sites have been identified in *rh1* promoter region suggesting that Otd may indirectly de-repress *rh1* in the inner PRs [108].

6.12   Specification of Photoreceptors in the Dorsal Rim Area (DRA)

The dorsal half of the eye contains specialized ommatidial cells that are involved in detecting polarized light. The dorsal rim area (DRA) consists of one to two rows of ommatidia that are directly adjacent to head cuticle (Fig. 6.11). The DRA PRs that are configured to sense polarized light depends on its strict alignment of microvilli in the rhabdomeres. The development of DRA ommatidia is controlled by Wg, which is expressed in the head cuticle surrounding the eye and its forced expression in the eye transform dorsal half of the ommatidia into DRA ommatidia. Additionally, Iroquois-complex (IRO-C) members (*auracan, caupolican,* and *mirror*) are also expressed in the dorsal half of the eye and forced expression of any of its member in the whole eye leads to the expansion of DRA in the ventral margin of the eye [109, 110]. Therefore, combinatorial action of Wg and IRO-C is required for DRA specification at the right place [110]. The homeodomain transcription factor Hth control DRA ommatidia specification in the eye [110]. Hth is expressed in R7 and R8 PRs of the DRA and its loss-of-function leads to an absence of DRA ommatidia in the eye that was confirmed by changes in rhabdomere morphology and loss of Rh3 in R7 and R8 (Fig. 6.11). Forced expression of Hth in all ommatidia is sufficient to transform them as DRA ommatidia, which are marked by increased inner PR diameter, expansion of Rh3 in all PRs and lack of normal coupling of Rh3/Rh5 and Rh4/Rh6 [110]. Interestingly, forced expression of Ss is not sufficient to induce Rh4 expression in the DRA ommatidia and therefore Hth is believed to antagonize Ss function in the DRA.

Questions
1. How does an ommatidium form and which molecular pathways are controlling it?
2. Describe the steps acting to determine cell fates and rhodopsin expression late in distinct PRs.

Take-Home Message
Ever since Darwin wrote the famous book on origin of species, there was a big debate to understand whether eyes evolved only once very early in the animal evolution and diversified thereafter (monophyletic origin) or if eyes evolved multiple times independently as evolution progressed (polyphyletic origin). The discovery of *Pax6* homolog *ey* in *Drosophila* and other RDN genes suggested homology and argued that all eyes are at least evolved from a common ancestral visual organ. It is now quite clear that the eye specification cascade is a conserved unit among all seeing animals and that several retinal disorders in humans can now be directly correlated in *Drosophila* and other species. Since *Drosophila* field has advanced genetic tools available, it provides an excellent system to screen new candidates in order to improve diagnosis and to understand eye-related disorders in more details. The recent advent of high-throughput assays such as single-cell RNA sequencing (scRNA seq) offers opportunity to detect subcellular changes and to provide new insights to understand eye development in precise details. Chromatin Immunoprecipitation coupled with next-generation sequencing (ChIP-seq) is an additional genomic technique that can be useful to find molecular interactors of RDN genes during retina development. Additionally, Gal4/UAS system in *Drosophila* could be routinely used for screening new genes involved in eye determination process in a spatiotemporal manner. In future, the development of advance genetic, molecular, and biochemical tool in *Drosophila* would provide novel insights to understand visual system development in more detail.

References

1. von Salvini-Plawen L, Mayr E. On the evolution of photorecptors and eyes. Plenum Press; 1977.
2. Quiring R, Walldorf U, Kloter U, Gehring WJ. Homology of the eyeless gene of *Drosophila* to the Small eye gene in mice and Aniridia in humans. Science. 1994;265(5173):785–9.
3. Hill RE, Favor J, Hogan BL, Ton CC, Saunders GF, Hanson IM, et al. Mouse small eye results from mutations in a paired-like homeobox-containing gene. Nature. 1991;354(6354):522–5.
4. Ton CC, Hirvonen H, Miwa H, Weil MM, Monaghan P, Jordan T, et al. Positional cloning and characterization of a paired box- and homeobox-containing gene from the aniridia region. Cell. 1991;67(6):1059–74.
5. Halder G, Callaerts P, Gehring WJ. Induction of ectopic eyes by targeted expression of the *eyeless* gene in *Drosophila*. Science. 1995;267(5205):1788–92.
6. Callaerts P, Munoz-Marmol AM, Glardon S, Castillo E, Sun H, Li WH, et al. Isolation and expression of a Pax-6 gene in the regenerating and intact Planarian Dugesia(G)tigrina. Proc Natl Acad Sci USA. 1999;96(2):558–63.
7. Glardon S, Callaerts P, Halder G, Gehring WJ. Conservation of *Pax-6* in a lower chordate, the ascidian *Phallusia mammillata*. Development. 1997;124(4):817–25.
8. Glardon S, Holland LZ, Gehring WJ, Holland ND. Isolation and developmental expression of the amphioxus Pax-6 gene (AmphiPax-6): insights into eye and photoreceptor evolution. Development. 1998;125(14):2701–10.
9. Halder G, Callaerts P, Gehring WJ. New perspectives on eye evolution. Curr Opin Genet Dev. 1995;5(5):602–9.
10. Nishina S, Kohsaka S, Yamaguchi Y, Handa H, Kawakami A, Fujisawa H, et al. PAX6 expression in the developing human eye. Br J Ophthalmol. 1999;83(6):723–7.
11. Onuma Y, Takahashi S, Asashima M, Kurata S, Gehring WJ. Conservation of Pax 6 function and upstream activation by Notch signaling in eye development of frogs and flies. Proc Natl Acad Sci USA. 2002;99(4):2020–5.
12. Strickler AG, Yamamoto Y, Jeffery WR. Early and late changes in Pax6 expression accompany eye degeneration during cavefish development. Dev Genes Evol. 2001;211(3):138–44.
13. Terzic J, Saraga-Babic M. Expression pattern of PAX3 and PAX6 genes during human embryogenesis. Int J Dev Biol. 1999;43(6):501–8.
14. Tomarev SI, Callaerts P, Kos L, Zinovieva R, Halder G, Gehring W, et al. Squid Pax-6 and eye development. Proc Natl Acad Sci USA. 1997;94(6):2421–6.
15. Tsachaki M, Sprecher SG. Genetic and developmental mechanisms underlying the formation of the *Drosophila* compound eye. Dev Dyn. 2012;241(1):40–56.
16. Kumar JP, Moses K. Expression of evolutionarily conserved eye specification genes during *Drosophila* embryogenesis. Dev Genes Evol. 2001;211(8–9):406–14.
17. Treisman JE, Heberlein U. Eye development in *Drosophila*: formation of the eye field and control of differentiation. Curr Top Dev Biol. 1998;39:119–58.
18. Ready DF, Hanson TE, Benzer S. Development of the *Drosophila* retina, a neurocrystalline lattice. Dev Biol. 1976;53(2):217–40.
19. Hardie RC. Functional organization of the fly retina. Progress in sensory physiology. Springer. 1985:1–79.
20. O'Tousa JE, Baehr W, Martin RL, Hirsh J, Pak WL, Applebury ML. The *Drosophila ninaE* gene encodes an opsin. Cell. 1985;40(4):839–50.
21. Zuker CS, Cowman AF, Rubin GM. Isolation and structure of a rhodopsin gene from *D. melanogaster*. Cell. 1985;40(4):851–8.

22. Sprecher SG, Desplan C. Chapter 4 – Development of the *Drosophila* melanogaster eye: from precursor specification to terminal differentiation. In: Tsonis PA, editor. Animal models in eye research. London: Academic Press; 2008. p. 27–47.

23. Gao S, Takemura SY, Ting CY, Huang S, Lu Z, Luan H, et al. The neural substrate of spectral preference in *Drosophila*. Neuron. 2008;60(2):328–42.

24. Yamaguchi S, Desplan C, Heisenberg M. Contribution of photoreceptor subtypes to spectral wavelength preference in *Drosophila*. Proc Natl Acad Sci USA. 2010;107(12):5634–9.

25. Wolff TRDF. Pattern formation in the Drosophila retina. The development of *Drosophila* melanogaster 1993:1277–1325.

26. Bonini NM, Leiserson WM, Benzer S. The *eyes absent* gene: genetic control of cell survival and differentiation in the developing *Drosophila* eye. Cell. 1993;72(3):379–95.

27. Braid LR, Verheyen EM. Drosophila nemo promotes eye specification directed by the retinal determination gene network. Genetics. 2008;180(1):283–99.

28. Curtiss J, Burnett M, Mlodzik M. *Distal antenna* and *distal antenna-related* function in the retinal determination network during eye development in *Drosophila*. Dev Biol. 2007;306(2):685–702.

29. Mardon G, Solomon NM, Rubin GM. *Dachshund* encodes a nuclear protein required for normal eye and leg development in *Drosophila*. Development. 1994;120(12):3473–86.

30. Serikaku MA, O'Tousa JE. *Sine oculis* is a homeobox gene required for *Drosophila* visual system development. Genetics. 1994;138(4):1137–50.

31. Singh A, Kango-Singh M, Sun YH. Eye suppression, a novel function of *teashirt*, requires Wingless signaling. Development. 2002;129(18):4271–80.

32. Haynie JL, Bryant PJ. Development of the eye-antenna imaginal disc and morphogenesis of the adult head in *Drosophila melanogaster*. J Exp Zool. 1986;237(3):293–308.

33. Kumar JP, Moses K. EGF receptor and Notch signaling act upstream of Eyeless/Pax6 to control eye specification. Cell. 2001;104(5):687–97.

34. Mishra AK, Sprecher SG. Early eye development: specification and determination. In: Molecular genetics of axial patterning, growth and disease in *Drosophila* eye. Springer; 2020. p. 1–52.

35. Hanson IM. Mammalian homologues of the *Drosophila* eye specification genes. Semin Cell Dev Biol. 2001;12(6):475–84.

36. Aldaz S, Morata G, Azpiazu N. The Pax-homeobox gene eyegone is involved in the subdivision of the thorax of *Drosophila*. Development. 2003;130(18):4473–82.

37. Cheyette BN, Green PJ, Martin K, Garren H, Hartenstein V, Zipursky SL. The *Drosophila sine oculis* locus encodes a homeodomain-containing protein required for the development of the entire visual system. Neuron. 1994;12(5):977–96.

38. Choi KW, Benzer S. Rotation of photoreceptor clusters in the developing *Drosophila* eye requires the *nemo* gene. Cell. 1994;78(1):125–36.

39. Czerny T, Halder G, Kloter U, Souabni A, Gehring WJ, Busslinger M. *Twin of eyeless*, a second *Pax-6* gene of *Drosophila*, acts upstream of *eyeless* in the control of eye development. Mol Cell. 1999;3(3):297–307.

40. Jun S, Wallen RV, Goriely A, Kalionis B, Desplan C. Lune/eye gone, a Pax-like protein, uses a partial paired domain and a homeodomain for DNA recognition. Proc Natl Acad Sci USA. 1998;95(23):13720–5.

41. Laugier E, Yang Z, Fasano L, Kerridge S, Vola C. A critical role of *teashirt* for patterning the ventral epidermis is masked by ectopic expression of *tiptop*, a paralog of *teashirt* in *Drosophila*. Dev Biol. 2005;283(2):446–58.

42. Pai CY, Kuo TS, Jaw TJ, Kurant E, Chen CT, Bessarab DA, et al. The Homothorax homeoprotein activates the nuclear localization of another homeoprotein, extradenticle, and suppresses eye development in *Drosophila*. Genes Dev. 1998;12(3):435–46.

43. Pan D, Rubin GM. Targeted expression of *teashirt* induces ectopic eyes in *Drosophila*. Proc Natl Acad Sci USA. 1998;95(26):15508–12.
44. Seimiya M, Gehring WJ. The *Drosophila* homeobox gene optix is capable of inducing ectopic eyes by an eyeless-independent mechanism. Development. 2000;127(9):1879–86.
45. Kumar JP. The molecular circuitry governing retinal determination. Biochim Biophys Acta. 2009;1789(4):306–14.
46. Kumar JP. Retinal determination the beginning of eye development. Curr Top Dev Biol. 2010;93:1–28.
47. Baonza A, Freeman M. Control of *Drosophila* eye specification by Wingless signalling. Development. 2002;129(23):5313–22.
48. Chen R, Halder G, Zhang Z, Mardon G. Signaling by the TGF-beta homolog *decapentaplegic* functions reiteratively within the network of genes controlling retinal cell fate determination in *Drosophila*. Development. 1999;126(5):935–43.
49. Hsiao FC, Williams A, Davies EL, Rebay I. Eyes absent mediates cross-talk between retinal determination genes and the receptor tyrosine kinase signaling pathway. Dev Cell. 2001;1(1):51–61.
50. Kurata S, Go MJ, Artavanis-Tsakonas S, Gehring WJ. Notch signaling and the determination of appendage identity. Proc Natl Acad Sci USA. 2000;97(5):2117–22.
51. Treisman JE, Rubin GM. wingless inhibits morphogenetic furrow movement in the *Drosophila* eye disc. Development. 1995;121(11):3519–27.
52. Bessa J, Carmona L, Casares F. Zinc-finger paralogues tsh and tio are functionally equivalent during imaginal development in *Drosophila* and maintain their expression levels through auto- and cross-negative feedback loops. Dev Dyn. 2009;238(1):19–28.
53. Datta RR, Lurye JM, Kumar JP. Restriction of ectopic eye formation by *Drosophila teashirt* and *tiptop* to the developing antenna. Dev Dyn. 2009;238(9):2202–10.
54. Dominguez M, Ferres-Marco D, Gutierrez-Avino FJ, Speicher SA, Beneyto M. Growth and specification of the eye are controlled independently by Eyegone and Eyeless in *Drosophila melanogaster*. Nat Genet. 2004;36(1):31–9.
55. Tomlinson A, Ready DF. Neuronal differentiation in *Drosophila* ommatidium. Dev Biol. 1987;120(2):366–76.
56. Wolff T, Ready DF. The beginning of pattern formation in the *Drosophila* compound eye: the morphogenetic furrow and the second mitotic wave. Development. 1991;113(3):841–50.
57. Kumar JP. My what big eyes you have: how the *Drosophila* retina grows. Dev Neurobiol. 2011;71(12):1133–52.
58. Bessa J, Gebelein B, Pichaud F, Casares F, Mann RS. Combinatorial control of *Drosophila* eye development by eyeless, homothorax, and teashirt. Genes Dev. 2002;16(18):2415–27.
59. Pichaud F, Casares F. *homothorax* and *iroquois-C* genes are required for the establishment of territories within the developing eye disc. Mech Dev. 2000;96(1):15–25.
60. Blochlinger K, Jan LY, Jan YN. Postembryonic patterns of expression of cut, a locus regulating sensory organ identity in *Drosophila*. Development. 1993;117(2):441–50.
61. Lebovitz RM, Ready DF. Ommatidial development in *Drosophila* eye disc fragments. Dev Biol. 1986;117(2):663–71.
62. Pignoni F, Hu B, Zavitz KH, Xiao J, Garrity PA, Zipursky SL. The eye-specification proteins So and Eya form a complex and regulate multiple steps in *Drosophila* eye development. Cell. 1997;91(7):881–91.
63. Salzer CL, Kumar JP. Position dependent responses to discontinuities in the retinal determination network. Dev Biol. 2009;326(1):121–30.
64. Horsfield J, Penton A, Secombe J, Hoffman FM, Richardson H. *decapentaplegic* is required for arrest in G1 phase during Drosophila eye development. Development. 1998;125(24):5069–78.

65. Chanut F, Heberlein U. Role of *decapentaplegic* in initiation and progression of the morphogenetic furrow in the developing *Drosophila* retina. Development. 1997;124(2):559–67.
66. Curtiss J, Mlodzik M. Morphogenetic furrow initiation and progression during eye development in *Drosophila*: the roles of *decapentaplegic, hedgehog and eyes absent*. Development. 2000;127(6):1325–36.
67. Greenwood S, Struhl G. Progression of the morphogenetic furrow in the *Drosophila* eye: the roles of Hedgehog, Decapentaplegic and the Raf pathway. Development. 1999;126(24):5795–808.
68. Kumar JP, Moses K. The EGF receptor and notch signaling pathways control the initiation of the morphogenetic furrow during *Drosophila* eye development. Development. 2001;128(14):2689–97.
69. Ma C, Zhou Y, Beachy PA, Moses K. The segment polarity gene hedgehog is required for progression of the morphogenetic furrow in the developing *Drosophila* eye. Cell. 1993;75(5):927–38.
70. Dominguez M, Hafen E. Hedgehog directly controls initiation and propagation of retinal differentiation in the *Drosophila* eye. Genes Dev. 1997;11(23):3254–64.
71. Frankfort BJ, Mardon G. R8 development in the *Drosophila* eye: a paradigm for neural selection and differentiation. Development. 2002;129(6):1295–306.
72. Hsiung F, Moses K. Retinal development in *Drosophila*: specifying the first neuron. Hum Mol Genet. 2002;11(10):1207–14.
73. Dominguez M. Dual role for Hedgehog in the regulation of the proneural gene *atonal* during ommatidia development. Development. 1999;126(11):2345–53.
74. Frankfort BJ, Nolo R, Zhang Z, Bellen H, Mardon G. *senseless* repression of rough is required for R8 photoreceptor differentiation in the developing *Drosophila* eye. Neuron. 2001;32(3):403–14.
75. Freeman M. Reiterative use of the EGF receptor triggers differentiation of all cell types in the *Drosophila* eye. Cell. 1996;87(4):651–60.
76. Freeman M. Cell determination strategies in the *Drosophila* eye. Development. 1997;124(2):261–70.
77. Kumar JP, Tio M, Hsiung F, Akopyan S, Gabay L, Seger R, et al. Dissecting the roles of the *Drosophila* EGF receptor in eye development and MAP kinase activation. Development. 1998;125(19):3875–85.
78. Yang L, Baker NE. Role of the EGFR/Ras/Raf pathway in specification of photoreceptor cells in the *Drosophila* retina. Development. 2001;128(7):1183–91.
79. Tomlinson A, Kimmel BE, Rubin GM. *Rough*, a *Drosophila* homeobox gene required in photoreceptors R2 and R5 for inductive interactions in the developing eye. Cell. 1988;55(5):771–84.
80. Kramer S, West SR, Hiromi Y. Cell fate control in the *Drosophila* retina by the orphan receptor seven-up: its role in the decisions mediated by the ras signaling pathway. Development. 1995;121(5):1361–72.
81. Dickson B, Hafen E. The development of *Drosophila* 1993.
82. Heberlein U, Mlodzik M, Rubin GM. Cell-fate determination in the developing *Drosophila* eye: role of the rough gene. Development. 1991;112(3):703–12.
83. Mlodzik M, Hiromi Y, Weber U, Goodman CS, Rubin GM. The *Drosophila* seven-up gene, a member of the steroid receptor gene superfamily, controls photoreceptor cell fates. Cell. 1990;60(2):211–24.
84. Domingos PM, Mlodzik M, Mendes CS, Brown S, Steller H, Mollereau B. Spalt transcription factors are required for R3/R4 specification and establishment of planar cell polarity in the *Drosophila* eye. Development. 2004;131(22):5695–702.

85. Daga A, Karlovich CA, Dumstrei K, Banerjee U. Patterning of cells in the *Drosophila* eye by Lozenge, which shares homologous domains with AML1. Genes Dev. 1996;10(10):1194–205.

86. Higashijima S, Kojima T, Michiue T, Ishimaru S, Emori Y, Saigo K. Dual Bar homeo box genes of *Drosophila* required in two photoreceptor cells, R1 and R6, and primary pigment cells for normal eye development. Genes Dev. 1992;6(1):50–60.

87. Tomlinson A, Bowtell DD, Hafen E, Rubin GM. Localization of the sevenless protein, a putative receptor for positional information, in the eye imaginal disc of *Drosophila*. Cell. 1987;51(1):143–50.

88. Basler K, Christen B, Hafen E. Ligand-independent activation of the sevenless receptor tyrosine kinase changes the fate of cells in the developing *Drosophila* eye. Cell. 1991;64(6):1069–81.

89. Begemann G, Michon AM, vd Voorn L, Wepf R, Mlodzik M. The *Drosophila* orphan nuclear receptor seven-up requires the Ras pathway for its function in photoreceptor determination. Development. 1995;121(1):225–35.

90. Hiromi Y, Mlodzik M, West SR, Rubin GM, Goodman CS. Ectopic expression of seven-up causes cell fate changes during ommatidial assembly. Development. 1993;118(4):1123–35.

91. Cooper MT, Bray SJ. Frizzled regulation of Notch signalling polarizes cell fate in the *Drosophila* eye. Nature. 1999;397(6719):526–30.

92. Tomlinson A, Struhl G. Delta/Notch and Boss/Sevenless signals act combinatorially to specify the *Drosophila* R7 photoreceptor. Mol Cell. 2001;7(3):487–95.

93. Cook T, Pichaud F, Sonneville R, Papatsenko D, Desplan C. Distinction between color photoreceptor cell fates is controlled by Prospero in *Drosophila*. Dev Cell. 2003;4(6):853–64.

94. Kauffmann RC, Li S, Gallagher PA, Zhang J, Carthew RW. Ras1 signaling and transcriptional competence in the R7 cell of *Drosophila*. Genes Dev. 1996;10(17):2167–78.

95. Kumar JP, Ready DF. Rhodopsin plays an essential structural role in *Drosophila* photoreceptor development. Development. 1995;121(12):4359–70.

96. Hardie RC, Raghu P. Visual transduction in *Drosophila*. Nature. 2001;413(6852):186–93.

97. Campos AR, Fischbach KF, Steller H. Survival of photoreceptor neurons in the compound eye of *Drosophila* depends on connections with the optic ganglia. Development. 1992;114(2):355–66.

98. Chang HY, Ready DF. Rescue of photoreceptor degeneration in rhodopsin-null *Drosophila* mutants by activated Rac1. Science. 2000;290(5498):1978–80.

99. Kirschfeld K, Franceschini N. Photostable pigments within the membrane of photoreceptors and their possible role. Biophys Struct Mech. 1977;3(2):191–4.

100. Morante J, Desplan C. Building a projection map for photoreceptor neurons in the *Drosophila* optic lobes. Semin Cell Dev Biol. 2004;15(1):137–43.

101. Mikeladze-Dvali T, Desplan C, Pistillo D. Flipping coins in the fly retina. Curr Top Dev Biol. 2005;69:1–15.

102. Wernet MF, Desplan C. Building a retinal mosaic: cell-fate decision in the fly eye. Trends Cell Biol. 2004;14(10):576–84.

103. Fortini ME, Rubin GM. The optic lobe projection pattern of polarization-sensitive photoreceptor cells in *Drosophila melanogaster*. Cell Tissue Res. 1991;265(1):185–91.

104. Mollereau B, Dominguez M, Webel R, Colley NJ, Keung B, de Celis JF, et al. Two-step process for photoreceptor formation in *Drosophila*. Nature. 2001;412(6850):911–3.

105. Wernet MF, Mazzoni EO, Celik A, Duncan DM, Duncan I, Desplan C. Stochastic spineless expression creates the retinal mosaic for colour vision. Nature. 2006;440(7081):174–80.

106. Mikeladze-Dvali T, Wernet MF, Pistillo D, Mazzoni EO, Teleman AA, Chen YW, et al. The growth regulators warts/lats and melted interact in a bistable loop to specify opposite fates in *Drosophila* R8 photoreceptors. Cell. 2005;122(5):775–87.

107. Vandendries ER, Johnson D, Reinke R. orthodenticle is required for photoreceptor cell development in the *Drosophila* eye. Dev Biol. 1996;173(1):243–55.
108. Tahayato A, Sonneville R, Pichaud F, Wernet MF, Papatsenko D, Beaufils P, et al. Otd/Crx, a dual regulator for the specification of ommatidia subtypes in the *Drosophila* retina. Dev Cell. 2003;5(3):391–402.
109. Tomlinson A. Patterning the peripheral retina of the fly: decoding a gradient. Dev Cell. 2003;5(5):799–809.
110. Wernet MF, Labhart T, Baumann F, Mazzoni EO, Pichaud F, Desplan C. Homothorax switches function of *Drosophila* photoreceptors from color to polarized light sensors. Cell. 2003;115(3):267–79.

Neurogenetics of Memory, Learning, and Forgetting

7

Lucia de Andres-Bragado, Jenifer C. Kaldun, Simon G. Sprecher

What You Will Learn in This Chapter

Animals are able to learn from past experiences and change their behavior. To understand the mechanisms underlying these learning and memory processes, we need to study them within the framework of neurogenetics; combining anatomical, behavioral, and genetic tools. In this chapter, we will explore the changes that occur in the brain when memories are formed and stored. We will take a reductionist approach—using simple model organisms such as *Drosophila melanogaster* or *Aplysia*—to examine the neuronal and molecular basis of how animals learn and how memories are stored. Finally, we will explore the neuronal and molecular mechanisms necessary for the complementary process: forgetting.

7.1 Learning Is Essential for Survival

Efficiently acquiring and storing memories are paramount for the survival of numerous species. Avoiding predators or remembering places for shelter offer clear advantages—and require the same processes: being able to learn and to remember.

Learning relies on acquiring new information through our senses, and, most importantly, on storing it using molecular and cellular changes so that this information can be

L. de Andres-Bragado · J. C. Kaldun · S. G. Sprecher (✉)
Department of Biology, University of Fribourg, Fribourg, Switzerland
e-mail: lucia.deandres@frontiersin.org; jenifer.kaldun@unifr.ch; simon.sprecher@unifr.ch

© Springer Nature Switzerland AG 2023   129
B. Egger (ed.), *Neurogenetics*, Learning Materials in Biosciences,
https://doi.org/10.1007/978-3-031-07793-7_7

recalled in the future. There are multiple ways of classifying the types of learning, and, in this section, we will mostly focus on two: non-associative and associative learning.

7.1.1 Non-associative Learning

In non-associative learning, the repetition of a single stimulus can trigger changes in behavior. This form of learning can have two outcomes known as *habituation* or *sensitization* [1].

7.1.1.1 Two Types of Non-associative Learning

Habituation occurs when animals stop reacting to a novel and repeated stimulus—such as when humans get used to sleeping despite the traffic noise in the street nearby [2]. This stimulus is classified as neither dangerous nor advantageous, and—after a prolonged exposure—we stop reacting to it.

Sensitization is the opposite reaction; after repeated exposure to the stimulus, the response is amplified [3]. This is the case of an allergic sensitization, where subsequent exposures to a stimulus might lead to a more dramatic reaction. In this situation, our body intensifies the response to a stimulus that might have an essential meaning for survival.

7.1.1.2 Insights from Humans and Model Organisms: How Are Memories Stored in Nerve Cells?

A great deal of our current understanding of learning and memory comes from studying "simpler" model organisms (with fewer neurons than humans) such as *Drosophila*, *Aplysia*, rats or mice.

However, the first insights of the mechanisms responsible for learning and memory were obtained studying human patients. These early studies on learning were either non-invasive experiments—for instance, as asking individuals to remember lists of words—or analysis of patients' cognitive recovery after uncommon injuries or surgeries.

A landmark experiment, crucial to establishing the learning and memory discipline, was conducted during the late 1950s, when doctors removed both the hippocampus and the temporal lobe of an epileptic patient called Henry Molaison (also known as H.M.) [4]. After this operation, H.M. retained his personality and most of his cognitive abilities, with one clear exception: his ability to consolidate long-term memories (see Sect. 7.3.1.1). The outcome of this operation highlighted the importance of the hippocampus in learning and memory.

Since the hippocampus is a brain region exclusively found in vertebrates, this led to many researchers—including Nobel-prize-winner Eric Kandel—to turn to vertebrate models to study neuronal activity in their hippocampus with the aim of understanding the underlying processes behind learning and memory.

Soon afterward, it was proven that the basis of learning and memory is not exclusive to the hippocampus and that it relies heavily on the properties of neurons. However, since

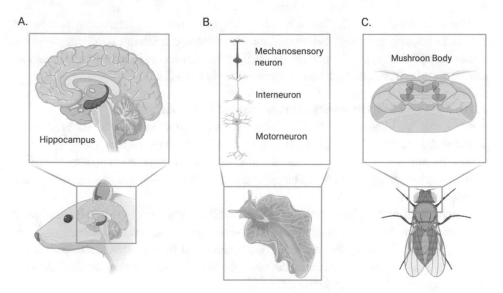

Fig. 7.1 Learning and memory centers in rats (**a**, hippocampus), in *Aplysia* (**b**, simple nervous circuit) and in *Drosophila* (Mushroom Body). Figure created with https://www.BioRender.com

most neurons share similar properties regardless of their location within the brain, it was hard to explain why learning takes place in the hippocampus and not in another region. This led to the hypothesis that the key difference to explain learning and memory was not based on the *location* of the neurons, but on something different: the changes in *interconnections* between them.

Since hippocampus in vertebrates have a high number of neurons—and therefore a lot of possible connections—it was not the right place to start, as monitoring all these neurons was not technically feasible. Solving the problem required taking a reductionist approach; finding an organism with significantly fewer neurons and interconnections, such as *Aplysia* (Fig. 7.1) [5].

7.1.1.3 Simple Behaviors can be Studied in *Aplysia*

Aplysia, a marine snail, belongs to mollusks that exhibit simple reflex behaviors like withdrawing its gill or siphon to defend itself.

Until Nobel prize laureate Eric Kandel started working on *Aplysia*, invertebrate model organisms were not used to study simple behaviors, let alone more complex ones such as learning. Studies in non-mammals had been crucial for breakthroughs in neurobiology— such as discovering the action potential in the squid giant axon [6], detecting synaptic transmission in the frog synapses, or understanding the synaptic transmission in the Limulus eye. But, at this point, a lot of very well-known scientists believed that the mechanisms of learning and memory could only be understood by studying humans, or at least vertebrates.

However, Kandel proved that not only did *Aplysia* show an interesting number of behaviors but also that these behaviors could be modified through experiences—giving rise to the different types of learning described by the classical psychologists: habituation, sensitization, and even classical conditioning (See Sect. 7.1.2.4).

7.1.1.4 Habituation in Invertebrates: Insights from *Aplysia*

Aplysia was well-known for having robust reflexes: after mechanical stimulation, it would withdraw the gill and siphon. However, experiments where *Aplysia* was repeatedly exposed to mechanical stimuli showed that, after repeated stimulation, it no longer showed a gill or syphon withdrawal reflex. This change in behavior proved that invertebrates could *habituate* [7].

Due to the simple nervous system of *Aplysia*, all neurons involved in the withdrawal mechanism were known: a mechanoreceptor sensory neuron, several interneurons, and a motor neuron. Knowing the neuronal components and their interconnections had the advantage that the cause for habituation could be studied looking at changes within the wiring diagram and the individual neurons.

At that time, there were two possible hypotheses to explain the underlying mechanisms of learning: either the neuronal connections would be changed or the strength between the synapses would be modulated. It turned out to be the latter—as predicted by Ramon y Cajal in the late nineteenth century [8]. The neuronal wiring diagram was determined during development and remained invariable, while the *strength* and *effectiveness* of synaptic connectivity changed depending on the learned experiences.

On the one hand, in a non-habituated *Aplysia*, a mechanical stimulus in the siphon would be sensed by the mechanosensory neurons, which would release glutamate and generate fast excitatory postsynaptic potentials (EPSPs) in a population of interneurons. This would in turn excite the motor neurons and lead to the reflex withdrawal response.

On the other hand, in a habituated *Aplysia*, the amount of glutamate released from the sensory neurons would decrease. This means that the presynaptic neuron would release less synaptic vesicles, which would in turn excite fewer glutamate receptors in the postsynaptic neuron. This decreased amount of glutamate leads to a decreased number of EPSPs, which ultimately results in a decrease (or even lack) of reflex.

Long-term habituation eventually changes the strength of the synapses involved in this pathway. This can be seen in anatomical studies which show that long-term habituation causes a reduced number of synaptic contacts between the mechanosensory cell and motor neuron [9, 10].

The key to answer why learning and memory occurs in certain regions of the brain is that not all synapses can be easily modified through experience. Some synapses—like the ones in the gill-withdrawal reflex in *Aplysia* or within the hippocampus in vertebrates—can considerably change their strength after just a few training sessions. Therefore, these areas can be considered the *learning centers*, the regions where synapses are modified after experiences are learned.

7.1.1.5 Sensitization in Invertebrates: Insights from *Aplysia*

When *Aplysia* are presented with harmful stimuli (such as an electric shock on the tail), they boost their reflex reaction trying to escape from this unwanted stimulus.

The neurons that modulate this enhanced response during sensitization are the same ones that modulate a decreased response during habituation. But, in this case, instead of a decrease of synaptic transmission, there is an increase.

This increase of synaptic transmission is controlled through several interneurons, one of the best-studied ones being the serotonergic ones. The release of serotonin from these interneurons leads to a series of downstream signaling cascades which ultimately promote the neurotransmitter release through two main mechanisms.

On the one hand, the protein kinase known as PKA phosphorylates the K^+ channels, closing them. This results in a broader action potential and an increase in Ca^{2+} influx through the voltage-gated calcium channels, which eventually leads to an increase in neurotransmitter release.

On the other hand, the protein kinase known as PKC phosphorylates a number of molecules that directly lead to an increased synaptic vesicle release (Fig. 7.2).

Both habituation and sensitization can be either short-term or long-term depending on the number of times that the initial stimulus is presented. A single electric shock would produce a short-term effect which would last for minutes, while at least five electric shocks would produce a long-term effect which would last up to weeks [3, 11–13].

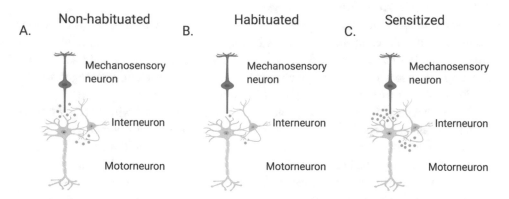

Fig. 7.2 The gill-withdrawal of *Aplysia* can be modulated depending on whether the animal is not habituated (**a**, left), habituated (**b**, middle), or sensitized (**c**, right). The simple sensory neuron–interneuron–motor neuron circuit in *Aplysia* is characterized by a different amount of neurotransmitter release in each of these situations, different amounts of neurotransmitters in each of these situations, and in *Drosophila* (Mushroom Body). Figure created with https://www.BioRender.com

7.1.2 Associative Learning

Associative learning is a more complicated process than habituation or sensitization. Instead of changing the behavior in response to a single stimulus, the animal must be presented two stimuli: a *conditioned* and an *unconditioned* stimulus.

The conditioned stimulus is one to which the animal's response is initially neutral to, one which does not trigger any specific response. Conversely, the unconditioned stimulus triggers a strong response. Therefore, unlike in habituation and sensitization, the animal integrates two stimuli [14].

7.1.2.1 Pavlov's Associative Learning Experiments

Associative learning was initially explored by Ivan Pavlov at the beginning of the twentieth century with his famous dog experiments. He identified a stimulus that dogs were initially be neutral to—a ringing bell—and a stimulus that triggered a strong and quantifiable behavioral response—exposure to meat. Then, he performed experiments where he rang the bell (conditioned stimulus) just before giving the dogs some meat (unconditioned stimulus). Pavlov observed that, after some training sessions pairing the conditioned and unconditioned stimulus, the dogs were able to predict the presence of meat. Whenever the bell started ringing, the dogs would start salivating, anticipating that the bell would be followed by meat—even if they were not given any meat in the end [15].

7.1.2.2 Associative Learning in *Aplysia*

Pavlov's principles of classical conditioning operate across a variety of behavioral responses and in different species, such as in *Aplysia*. Not only can *Aplysia* modulate the withdrawal reflex through sensitization and habituation but also by classical conditioning. However, behaviors learned through classical conditioning are not only stronger but also last longer than those acquired through sensitization and habituation.

When touching the siphon softly (conditioned stimulus) is paired with an electric shock in the tail (unconditioned stimulus), *Aplysia* associates both events, which enhances its gill-withdrawal reflex. Similar to Pavlov's salivating dogs, the conditioned stimulus must precede the unconditioned stimulus and both stimuli must be given within a short time-frame—ideally around half a second. If this is the case, *Aplysia* is able to *predict* the presence of the unconditioned stimulus and withdraws its gill more dramatically, even in the absence of the electric shock.

The cellular basis of associative learning in *Aplysia* relies on a process called *synaptic facilitation*. This process depends on the coincidence detection of both sensory neurons from the conditioned and unconditioned stimulus. In the case of *Aplysia*, after the electric shock is detected by the sensory neurons, the interneurons release serotonin on the synapse between the mechanosensory neuron (which is the neuron that detects the mechanical stimulus in the siphon) and the motor neuron. If the release of serotonin occurs right before the action potential from the mechanosensory neuron is triggered, this will lead to synaptic facilitation and ultimately an enhanced withdrawal of the gill [16].

7.1.2.3 Associative Learning in *Drosophila*

Associative learning is commonly studied in *Drosophila melanogaster* using different modalities of stimuli, such as odors, vision, or taste.

In the case of olfactory learning, flies are first given a conditional stimulus—an odor that the fly is initially neutral to—and this odor is repeatedly presented with either a reward (e.g., sugar) or a punishment (e.g., an electric shock). The fly is then presented with a second odor without any additional stimuli. After some training (either one session in the case of reward or several ones in the case of punishment) where both the conditional and unconditional stimulus are paired, the fly can freely decide between the two odors. Flies that have successfully learned will avoid or approach the odor previously paired with the unconditional stimulus. Most of the times, to be able to check if the flies are able to perceive the odors or not, slightly aversive odors are used [17, 18].

These experiments can be performed in the so-called Tully machine. This machine was designed by Tully and Quinn in the 1980s, as they wanted to test Pavlovian learning in *Drosophila* flies in a consistent and high-throughput way. This machine consists of a T-maze, where different odors are placed on either side of the maze, and flies are free to choose to go towards the paired stimulus, or in the opposite direction. Researchers have adapted and optimized this *Drosophila* learning protocol using the Tully machine to meet their different research questions.

An advantage of the Tully machine is that we can quantify flies' learning ability, calculated through a learning index. This index is the ratio between the flies which go towards the unconditioned odor and the second odor (preference index, Eq. 7.1). To avoid biases, flies are also trained with a second odor as the paired odor. The average of both reciprocal experiments defines the learning ability (learning index, Eq. 7.2) [17].

$$\text{Preference index} = 100 \times \frac{\text{Flies in paired odor} - \text{Flies control odor}}{\text{Total number of flies}} \tag{7.1}$$

$$\text{Learning index} = \frac{(\text{Preference 1} + \text{Preference 2})}{2} \tag{7.2}$$

Once we have calculated the flies' learning index under specific conditions (stimuli modality, or type of unconditioned stimulus) we can compare this learning index value between wild type flies and flies with other genotypes.

Due to the numerous genetic tools available in *Drosophila*, we can target both genes and neuronal circuits, gaining a comprehensive view of learning and memory mechanisms at a genetic, molecular, and cellular level (see Sect. 7.3.1.4).

7.1.2.4 Operant Conditioning: A Way of Associative Learning

Another type of associative learning is operant conditioning. Unlike classical conditioning—which involves an involuntary behavior learned after pairing stimuli with biologically relevant events—operant conditioning requires an active involvement of the animal. In operant conditioning, the animal forms an association between an action and its consequences.

A classical behavioral experiment used to study this type of learning is Skinner's model for operant conditioning, where a subject (such as a pigeon or a rat) is placed in a box where all of the stimuli are carefully controlled. This box contains a lever, and, if just by chance the rat presses it, food gets into the box. After some coincidental pressings of the lever, the rat learns that whenever it presses the lever, the food will get into the box and it starts *consciously* pressing the lever to get food. The animal has learnt the *consequences* of pressing the lever, and voluntary presses it to get more food [19].

7.2   Memory

After learning, animals need to be able to store and retrieve this information, they need to be able to *memorize* it. Memories can be classified according to different characteristics, like the amount of time for which these memories will be stored (long-term vs short-term), or the nature of these memories. A popular classification between psychologists is explicit or implicit memory—although sometimes memories cannot be unmistakably classified in just one of these two categories.

7.2.1   Explicit Memory

Explicit memory relics on *consciously* learning and *consciously* recalling, and it is a type of memory that is usually used for facts or concepts.

However, information that we initially store as explicit memory could be considered as implicit memory after some time. These types of memories are often linked to *visual* facts, or concepts but they can also be common facts that are very much present in our lives, such as the birthdays of family members.

Therefore, vertebrates are the most common animal models used to study the mechanisms of explicit memory.

7.2.2   Implicit Memory

Implicit memory is characterized by a lack of conscious awareness, and it is not linked to visual memories but rather to muscle memory—such as riding a bicycle.

On the other hand, the mechanisms underlying implicit memory formation can easily be studied using invertebrate animal models, such as the fruit fly *Drosophila melanogaster* or the European honey bee, *Apis mellifera*.

7.3   Mechanisms of Learning and Memory

To gain a better understanding of learning and memory processes, not only should we classify them into different types, but we should also understand the underlying processes, the *cellular* and *molecular* mechanisms.

7.3.1   Cellular Mechanisms

It was only when doctors started performing experimental surgeries to remove different parts of patient's brains (such as the removal of the hippocampus and the temporal lobe in epileptic patient Henry Molaison) that the location of the so-called learning centers was unveiled, and the cellular mechanisms of learning and memory could be studied.

7.3.1.1 The Case of Henry Molaison

In the case of patient H.M., an unfortunate accident in his childhood leads to him having periodic seizures and blackouts, which forced him to drop out of school.

Due to these inconvenient seizures, he agreed to go through a risky surgery, where important parts of his brain would have to be removed. Before his surgery was performed, the function of the hippocampus was completely unknown, but some people hypothesized that different mental capacities would be located in different brain areas.

After the operation, Henry Molaison seemed completely recovered: his seizures were nearly gone, and his personality remained intact. However, they soon discovered that the operation had not been completely successful, as he was unable to form new long-term memories.

Henry Molaison was able to do simple tasks, such as remembering a random number for up to fifteen minutes by constantly repeating it to himself, but if he would stop repeating it for five minutes, he would not only forget the random number but also the memory test itself.

The tests performed on Henry Molaison proved that there are different steps in memory formation: short-term memory and long-term memory. And, most importantly, that these two processes use different brain regions.

In humans, sensory data is temporarily transcribed by the neurons in the cortex. Next, this information is stored in the hippocampus, where specific proteins strengthen the cortical synaptic connections. If the memory is recalled periodically (or if it is a strong memory), the hippocampus transfers this information back to the cortex, where it can be "permanently" stored. Since Henry Molaison's hippocampus was removed, he was only able to retain memories for a short period of time.

7.3.1.2 The Hippocampus as a Learning Center

The difference between the hippocampus and other areas of the brain lies in its dynamic and fast neural plasticity, its ability to modulate the synaptic strength. In particular, a key form of neuronal plasticity known as long-term potentiation (LTP) was first discovered in the hippocampus and has afterwards been shown to take place in other areas of the brain including the sensorimotor synapse of *Aplysia*.

7.3.1.3 Long-Term Potentiation as a Mechanism to Store Memories

LTP is considered to be one of the main neural mechanisms by which memories are stored in the brain. In general, LTP is a widespread mechanism to enhance synaptic connections. Therefore, it depends on a strong activation of the pre- and postsynaptic neurons (through high-frequency action potentials).

Despite the lack of a universal process to induce LTP, a key requirement is the presence of a series of high-frequency nearly simultaneous action potentials, which should be targeted to this presynaptic neuron. This will ultimately strengthen a synapse that receives a double input from different presynaptic neurons.

As a result, an LTP will increase the size of the postsynaptic potentials within that synapse. This change in the postsynaptic potential size can last up to weeks, and proving as an efficient process to store memories.

The best described examples are the glutamatergic neurons in the hippocampus. At the molecular level, before the LTP potentiation takes place, the NMDA glutamate receptors in the postsynaptic neuron (which would normally be opened in response to glutamate) is blocked by a Mg^{2+} ion. As synaptic activity reaches the nearby synapses, the membrane of the postsynaptic is depolarized, which leads to the release of the Mg^{2+}. Once the postsynaptic receptor is unblocked, it responds to glutamate and allows other ions (such as Na^+ or Ca^{2+}) to go through, which leads to a further depolarization of the membrane. Furthermore, the influx of Ca^{2+} triggers the stored AMPA glutamate receptors to translocate to the postsynaptic membrane, allowing even more glutamate to go through.

In a synapse where an LTP has been established in the postsynaptic neuron, the AMPA receptors are already located in the postsynaptic membrane, and, therefore, the release of glutamate in the presynaptic neuron will directly trigger a depolarization. The combined action of both the AMPA and NMDA receptors located in the postsynaptic neuron trigger postsynaptic potentials which are strong enough to initiate action potentials without the input from other synapses.

7.3.1.4 *Drosophila*: A Model Organism to Map Neurons

Even though *Drosophila*, does not have the same brain regions as vertebrates, this model organism can help us understand the basic concepts involved in learning and memory, which can then be extrapolated to higher organisms.

Since *Drosophila* has fewer neurons than mammals, it is a key organism to study learning and memory both at the individual neuron and at the neuronal circuit level. This task is easier in *Drosophila* than in other organisms due to the great number of genetic tools available, which enables scientists to easily target individual neurons and trace down the circuits involved in different learning tasks, such as olfactory or visual learning.

7.3.1.5 *Drosophila's* Odor Learning Circuit

To obtain a complete map of the neuronal circuit involved in learning and memory, we need to understand how the information is processed between the sensory neurons— which are able to detect the conditional stimuli, such as odors—all the way to the motor neurons—which transmit this information to the muscles to execute the decisions. In between the sensory and the motor neurons, the information from the external stimuli must be integrated and associated in a brain structure where memories are formed.

In *Drosophila* we are at a point where we have mapped practically all the neuronal circuit involved in certain tasks, such as olfactory learning. In this case, odors are detected by the odor receptor neurons, which transmit this information to some interneurons and projection neurons within a brain region called the Antennal Lobe (AL). This information is then integrated in another brain region called the Mushroom Body (MB), where some dopaminergic neurons can reinforce or weaken the learnt experiences, depending on the associations made. Finally, the mushroom body output neurons are in charge of transmitting this information to the motor neurons. This process allows flies to make association between stimuli and show the correct motor output: either avoid or go towards a particular stimuli.

7.3.1.6 UAS-Gal4: A Binary Transcriptional System

One of the most widely used genetic tools used in *Drosophila* is the UAS-Gal4 system. This genetic tool is comprised of Gal4—a transcription factor present in yeast—and the Upstream Sequence Activator (or UAS)—a promotor sequence to which Gal4 can bind specifically. The UAS-Gal4 system is a binary system, as both components (UAS and Gal4) can be kept in independent fly libraries, and it can be combined in a subset of the offspring.

Using the UAS-Gal4 system, we can target gene expression both temporally (determining *when* it should be expressed) and spatially (determining *where* it should be expressed). To this end, the Gal4 transcription factor is inserted under the expression of a tissue-specific promotor. To ensure that the Gal4 protein leads to the expression of the desired gene, a specific gene of interest is placed under the expression of the UAS sequence. This gene of interest will only be expressed when Gal4 binds to it and drives its expression.

7.3.1.7 The Mushroom Body Is the Fly's Central Structure for Learning

The vast number of genetic tools available for *Drosophila melanogaster* had not been developed when scientists first started using it as a model for learning and memory, therefore, other strategies had to be used.

For instance, the first evidence showing that in *Drosophila* the Mushroom Body was the center for learning and memory was obtained by feeding fruit flies with hydroxyurea. This chemical selectively ablated the cells in the MB in a crucial developmental stage.

The lack of a part (or all) of the Mushroom Body did not have an effect on the flies' ability to sense different sensory stimuli, such as odors or electric shocks. However, it did have an effect on their ability to associate conditional stimuli (in this case odors) to electric shocks and learn to avoid them.

7.3.1.8 Shibire to Block Synaptic Transmission

With our current genetic tools, we can also ablate the mushroom body in *Drosophila* using the UAS-Gal4 system. For instance, the gene of interest expressed under the UAS sequence could be a gene that is required for learning or a gene that would prevent synaptic activity, such as *shibire^{ts}*, a temperature-sensitive dynamin mutant gene in *Drosophila melanogaster*. Activating *shibire* blocks endocytosis and prevents membrane recycling, which ultimately depletes the available synaptic vesicles.

There are also temperature-sensitive *shibire^{ts}* alleles, which can be temporally controlled (whenever the temperature is increased to 29 °C) and in a reversible manner (as reversing the temperature to 18 °C will stop the expression of *shibire^{ts}*). This *shibire^{ts}*-dependent inactivation of the MB proved that this structure is not only required, but that it is also necessary for learning.

7.3.1.9 The MB Synaptic Output Is Not Required During Memory Acquisition But It Is Necessary During Retrieval

Using a Mushroom-Body-specific Gal4 driver (such as mb247-Gal4), and designing the appropriate crosses, *shibire* can be specifically expressed only in the Mushroom Body.

The Gal4 system with *shibire* and MB-specific Gal4 drivers led to clean experiments which showed at what stages of the learning process the mushroom body was involved.

On the one hand, flies were first trained at a temperature where *shibire* would be active (and therefore none of the neurons in the mushroom body would be able to form synapses). Subsequently, the final training and the test were performed at a temperature where *shibire* would not be active. Such an experiment showed that flies without a functional Mushroom Body during the first phase of the learning process, during the memory acquisition, could form memories as efficiently as flies where the Mushroom Body was active during the whole period.

On the other hand, flies which were first trained at a temperature where *shibire* would be inactive and during the consolidation phase were transferred to a temperature where *shibire* would be active had much lower performance indexes than the flies with an intact Mushroom Body. Therefore, the Mushroom Body output is required for consolidation.

7.3.2   Molecular Mechanisms

Apart from deciphering the specific neurons and neuronal circuits involved in learning and memory, using model organisms we can also understand the underlying molecular processes.

7.3.2.1 Seymour Benzer: Identifying *Drosophila Melanogaster* Learning Mutants

An early way of understanding the molecular mechanisms involved in different processes was to use forward genetics to create thousands of mutants, which would subsequently be tested using straightforward behavioral tests such as, in the case of learning, odor learning assays. This is the approach that famous *Drosophilist* Seymour Benzer used to identify numerous mutants with different phenotypes.

Once a certain strain of flies with a defect in learning was found, the mutations were mapped to try and find the mutant gene responsible for this phenotype. This led to the discovery of important learning and memory mutants in *Drosophila*, such as *rutabaga*, or *dunce*.

7.3.2.2 *Drosophila* Learning and Memory Genes

As mentioned previously, the olfactory pathway in *Drosophila* has been used to decipher learning a memory mechanisms at the neuronal, molecular, and gene level.

Drosophila perceives odors via olfactory receptors of the olfactory sensory neurons in the antennae, which project to the Antennal lobe (AL) [20]. Within the antennal lobe, interneurons reshape the odor information before projection neurons send it to the Mushroom body dendrites, where acetylcholine-dependent Ca^{2+}-channels open upon neuronal activity [21–24]. The Mushroom body receives other sensory information like mechanical stimuli or taste by octopaminergic and dopaminergic neurons, which in turn release dopamine and octopamine and activate G-Protein coupled receptors [25, 26]. Simultaneous activation of the MB neurons by odors and G-protein signaling activates the cAMP-PKA-CREB signaling pathway [27]. This pathway is highly conserved in learning and memory and which can be found in various species such as mice, humans, and *Drosophila*.

The parallel increase in G-protein signaling and Ca^{2+} is detected by an adenyl cyclase, which functions as a coincidence detector. In *Drosophila*, the coincidence detector is *rutabaga,* an important gene for learning [28, 29]. *rutabaga* encodes for an adenyl cyclase coincidence detector, which responds both to the increase in Ca2+ and G-Protein signaling. Rutabaga responds by elevating cAMP levels, which activate the protein kinase A (PKA). PKA phosphorylates other proteins required for learning and for establishing short-term memory [29]. The main target of PKA is cAMP response element-binding protein (CREB), a transcription factor which regulates the expression of various memory genes. Without CREB-dependent translation transcription, short-term memories cannot be consolidated into long-term memories [27]. The levels of cAMP can be regulated via Dunce, a phosphodiesterase which inactivates cAMP and, therefore, inhibits learning [30].

Another important group of genes is involved in translation control, as the synthesis of new proteins is a key step to form long-term memories [31]. Since the synapses are far from the nucleus, and the transport of protein is too slow to respond dynamically to neuronal signaling, synapses have a pool of mRNA they can locally translate on demand. The mRNAs are transported by RNA binding proteins—which protect the RNA from being degraded—in ribonucleoparticles from the soma to dendrites and axonal termini. These mRNAs are kept transcriptionally inactive until required using microRNAs, which bind to it [32].

Many more genes were discovered in *Drosophila* to be involved in learning and memory. For example, cell adhesion molecules (like FasII) are involved in shaping the synaptic connection between neurons [33]. Another group of involved genes is the neurotransmitter receptors themselves, which allows to adapt their amount dynamically [34]. Another well-known learning gene in the flies is *amnesiac*, a neuropeptide involved in consolidation of memories [35].

All in all, learning is a balance between genes promoting learning and genes inhibiting it, and it is regulated at different levels, including the protein properties (via post-translational modification) and abundance (via transcription and translation control).

7.3.2.3 Molecular Mechanisms in *Aplysia*: Short-Term and Long-Term Sensitization

Studying other model organisms, such as *Aplysia* has also given us an enormous insight into the way that learning and memory occurs at a molecular level. In particular, studies in *Aplysia* have shown the molecular differences between the molecular processes that take place during short-term and long-term memory.

In general, short-term memory relies on a transient phosphorylation of ion channels, which leads to an increased release of neurotransmitter.

On the other hand, for long-term memories to be formed, a stronger and longer-lasting stimulus is required. This sustained stimulus leads to an increase in the level of the messenger molecule cAMP, which activates the downstream protein kinases. These kinases will not only phosphorylate different proteins, but they will also lead to the synthesis of new proteins. Ultimately, all these molecular changes will change the form and function of the synapse, which will gain a higher efficiency and increase its neurotransmitter release.

7.3.2.4 Enhancement Synaptic Efficacy

All in all, there are a number of cellular and molecular changes that can enhance synaptic efficacy, both in the presynaptic or in the postsynaptic neuron, or both.

Changes in the presynaptic neuron would include an increased probability of releasing synaptic vesicles, an increased number of versicle release sites, or an increased number of vesicles available to be released.

Changes in the postsynaptic neuron would be comprised of an increased sensitivity of the receptors, or an increase in the number of functional receptors.

Moreover, the connection between the neurons is adapted by modifying the adhesive molecules in between the cells.

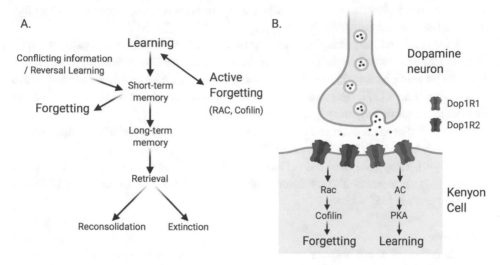

Fig. 7.3 (**a**) Information processing: Learned information is stored as short-term memory and consolidated as long-term memory. Upon retrieval, the information is either re-consolidated or forgotten. Memory formation is also challenged at the short-term memory stage by conflicting information. Importantly, learning activates competing forgetting pathways. (**b**) Forgetting pathway: In *Drosophila*, dopamine release triggers memory acquisition by signaling through dopamine receptor dop1R1 and the downstream AC-PKA-Creb signaling pathway. Dopamine also binds to dopamine receptor dop1R2, which is upstream of Rac and cofilin and therefore initiates forgetting

7.4   Forgetting

In the previous sections we learned how memories are formed and stored. In order to learn, we must consolidate some memories, and forget others. Therefore, forgetting is a crucial process for a normal brain function, and it is essential for learning and memory.

A way of classifying forgetting is by dividing it into passive and active forgetting. Passive forgetting describes the biological decay of memories over time, whereas active forgetting is triggered by interfering with memories during retrieval or by reversal learning. Recent studies suggest that active forgetting is an important function to retain normal brain function (Fig. 7.3).

7.4.1   Active Forgetting Pathways

Studies in fruit flies discovered an important mediator in active forgetting: the GTPase Rac, which is also involved in cytoskeleton dynamics. Using Gal80ts, a temperature-sensitive repressor of Gal4, the expression of Rac can be controlled to obtain either a constitutive active Rac expression or dominant negative Rac expression. At 18 °C—the permissive temperature—Gal80ts hinders activation of gene expression through Gal4, whereas at

30 °C—the restrictive temperature—Gal80ts is not functioning, therefore permitting Gal4 mediated gene expression [36].

On the one hand, inhibiting Rac activity by expressing a dominant negative isoform leads to slower memory decay. On the other hand, increasing Rac activity by expressing a constitutive active isoform causes faster memory decay [36]. Similar results were obtained in rat hippocampus [37], suggesting a conserved role for Rac and its downstream elements like cofilin in active forgetting.

Further studies in flies showed that learning-induced dopamine signaling is upstream of Rac, and an important factor in inducing forgetting [38, 39]. More and more players like scribble [40], sleep [41], and the GTPase Cdc42 [42, 43] as well as involved neuronal circuits [44] are described. This demonstrates that forgetting is a complex and highly regulated process [45].

Questions

1. What is the difference between associative and non-associative learning?
2. How would you test a fly mutant's ability to learn? What experiments would you perform?

Take-Home Message

This chapter describes the basic principles and mechanisms of learning, memory, and forgetting.

Learning can be differentiated into non-associative learning and associative learning, whereas memory is differentiated by the type of information or by the temporal aspect. Moreover, forgetting can be divided into passive and active forgetting.

Invertebrate models like *Aplysia* or *Drosophila* have given us tremendous insight into the molecular and cellular aspects of brain functions, such as the use of the cAMP-PKA-CREB pathway signaling pathway to encode long-lasting memories.

In recent years, it became obvious that forgetting is an integral part of efficient memory formation. It involves the small G-protein Rac and induces re-arrangements of the cytoskeleton.

References

1. Poon C-S, Schmid S. Nonassociative learning, Seel NM, editor. Boston: Springer US; 2012. p. 2475–7
2. Tsang CD. Habituation. Seel NM, editor. Boston: Springer US; 2012. p. 1411–3
3. Pinsker HM, Hening WA, Carew TJ, Kandel ER. Long-term sensitization of a defensive withdrawal reflex in *Aplysia*. Science. 1973;182(4116):1039–42

4. Shah B, Pattanayak R, Sagar R. The study of patient Henry Molaison and what it taught us over past 50 years: contributions to neuroscience. J Ment Health Hum Behav. 2014;19(2):91–3

5. Kandel ER. Nobel lecture. https://www.nobelprize.org/prizes/medicine/2000/kandel/lecture/, 2000. Online; Accessed 3 Feb 2020

6. Hodgkin AL, Huxley AF. A quantitative description of membrane current and its application to conduction and excitation in nerve. J Physiol. 1952;117:500–44

7. Pinsker H, Kupfermann I, Castellucci V, Kandel E. Habituation and dishabituation of the gill-withdrawal reflex in *Aplysia*. Science 1970;167(3926):1740–2

8. Ramon y Cajal S, de Felipe J, Jones EG. Cajal's degeneration and regeneration of the nervous system. Oxford: Oxford University Press; 1991

9. Castellucci V, Pinsker H, Kupfermann I, Kandel ER. Neuronal mechanisms of habituation and dishabituation of the gill-withdrawal reflex in *Aplysia*. Science. 1970;167(3926):1745–8

10. Carew TJ, Pinsker HM, Kandel ER. Long-term habituation of a defensive withdrawal reflex in *Aplysia*. Science. 1972;175(4020):451–4

11. Klein M, Kandel ER. Presynaptic modulation of voltage-dependent ca2+ current: mechanism for behavioral sensitization in *Aplysia californica*. Proc Natl Acad Sci. 1978;75(7):3512–6

12. Zhou L, Baxter DA, Byrne JH. Contribution of pkc to the maintenance of 5-ht-induced short-term facilitation at sensorimotor synapses of *Aplysia*. J Neurophysiol. 2014;112(8):1936–49. PMID: 25031258.

13. Barbas D, DesGroseillers L, Castellucci VF, Carew TJ, Marinesco S. Multiple serotonergic mechanisms contributing to sensitization in *Aplysia*: evidence of diverse serotonin receptor subtypes. Learn Mem 2003;10(5):373–86

14. Jozefowiez J. Associative learning. Seel NM. Boston: Springer US; 2012, p. 330–4

15. Clark RE. The classical origins of pavlov's conditioning. Integr Psych Behav 2004;39:279–94

16. Carew TJ, Walters ET, Kandel ER. Associative learning in *Aplysia*: cellular correlates supporting a conditioned fear hypothesis. Science. 1981;211(4481):501–4

17. Tully T, Quinn WG. Classical conditioning and retention in normal and mutant *Drosophila melanogaster*. J Comp Physiol. 1985;157:263–77

18. Krashes MJ, Waddell S. *Drosophila* appetitive olfactory conditioning. Cold Spring Harb Protoc. 2011;2011(5):pdb.prot5609

19. Skinner BF. The behavior of organisms: an experimental analysis. Cambridge: B. F. Skinner Foundation; 2019

20. Vosshall LB, Wong AM, Axel R. An olfactory sensory map in the fly brain. Cell. 2000;102(2):147–59

21. Ng M, Roorda RD, Lima SQ, Zemelman BV, Morcillo P, Miesenböck G. Transmission of olfactory information between three populations of neurons in the antennal lobe of the fly. Neuron. 2002;36(3):463–74

22. Wilson RI, Turner GC, Laurent G. Transformation of olfactory representations in the *Drosophila* antennal lobe. Science 2004;303(5656):366–70

23. Tanaka NK, Awasaki T, Shimada T, Ito K. Integration of chemosensory pathways in the textitDrosophila second-order olfactory centers. Curr Biol 2004;14(6):449–57

24. Yasuyama K, Meinertzhagen IA, Schürmann F-W. Synaptic organization of the mushroom body calyx in *Drosophila melanogaster*. J Comp Neurol. 2002;445:211–26

25. Kim Y-C, Lee H-G, Han K-A. D1 dopamine receptor dda1 is required in the mushroom body neurons for aversive and appetitive learning in *Drosophila*. J Neurosci. 2007;27(29):7640–7

26. Sabandal JM, Sabandal PR, Kim Y-C, Han K-A. Concerted actions of octopamine and dopamine receptors drive olfactory learning. J Neurosci. 2020;40(21):4240–50

27. Yin JCP, Del Vecchio M, Zhou H, Tully T. CREB as a memory modulator: Induced expression of a dCREB2 activator isoform enhances long-term memory in *Drosophila*. Cell. 1995;81:107–15

28. Livingstone MS, Sziber PP, Quinn W. Loss of calcium/calmodulin responsiveness in adenylate cyclase of rutabaga, a drosophila learning mutant. Cell. 1984;37:205–15
29. Gervasi N, Tchenio P, Preat T. PKA dynamics in a drosophila learning center: coincidence detection by rutabaga adenylyl cyclase and spatial regulation by dunce phosphodiesterase. Neuron. 2010;65:516–29
30. Dudai Y, Jan YN, Byers D, Quinn WG, Benzer S. Dunce, a mutant of *Drosophila* deficient in learning. Proc Natl Acad Sci 1976;73(5):1684–88
31. Bourtchouladze R, Abel T, Berman N, Gordon R, Lapidus K, Kandel ER. Different training procedures recruit either one or two critical periods for contextual memory consolidation, each of which requires protein synthesis and PKA. Learn Mem 1998;5(4):365–74
32. Rajgor D, Welle TM, Smith KR. The coordination of local translation, membranous organelle trafficking, and synaptic plasticity in neurons. Front Cell Dev Biol. 2021;9:1776
33. Cheng YZ, Endo K, Wu K, Rodan AR, Heberlein U, Davis RL. *Drosophila* fasciclinII is required for the formation of odor memories and for normal sensitivity to alcohol. Cell. 2001;105:757–68
34. Crocker A, Guan X-J, Murphy CT, Murthy M. Cell-type-specific transcriptome analysis in the *Drosophila* mushroom body reveals memory-related changes in gene expression. Cell Rep. 2016;15:1580–96
35. Quinn WG, Sziber PP, Booker R. The *Drosophila* memory mutant amnesiac. Nature. 1979;277:212–4
36. Shuai Y, Lu B, Hu Y, Wang L, Sun K, Zhong Y. Forgetting is regulated through Rac activity in *Drosophila*. Cell. 2010;140(4):579–89
37. Liu Y, Du S, Lv L, Lei B, Shi W, Tang Y, Wang L, Zhong Y. Hippocampal activation of Rac1 regulates the forgetting of object recognition memory. Cell. 2016;26(17):2351–7
38. Berry JA, Cervantes-Sandoval I, Nicholas EP, Davis RL. Dopamine is required for learning and forgetting in *Drosophila*. Neuron. 2012;74:530–42
39. Berry JA, Phan A, Davis RL. Dopamine neurons mediate learning and forgetting through bidirectional modulation of a memory trace. Cell Rep. 2018;25:651–62
40. Cervantes-Sandoval I, Chakraborty M, MacMulien C, Davis RL. Scribble scaffolds a signalosome for active forgetting. Neuron. 2016;90:1230–42
41. Berry JA, Cervantes-Sandoval I, Chakraborlty M, Davis RL. Sleep facilitates memory by blocking dopamine neuron-mediated forgetting. Cell. 2015;161:1656–67
42. Zhang X, Li Q, Wang L, Liu ZJ, Zhong Y. Cdc42-dependent forgetting regulates repetition effect in prolonging memory retention. Cell Rep. 2016;16:817–25
43. Gao Y, Shuai Y, Zhang X, Peng Y, Wang L, He J, Zhong Y, Li Q. Genetic dissection of active forgetting in labile and consolidated memories in *Drosophila*. Proc Natl Acad Sci U S A. 2019;116:21191–7
44. Shuai Y, Hirokawa A, Ai Y, Zhang M, Li W, Zhong Y. Dissecting neural pathways for forgetting in *Drosophila* olfactory aversive memory. Proc Natl Acad Sci U S A. 2015;112:E6663–72
45. Medina JH. Neural, cellular and molecular mechanisms of active forgetting. Front Syst Neurosci. 2018;12:3

Evolution and Origins of Nervous Systems

8

Jules Duruz, Simon G. Sprecher

What You Will Learn in This Chapter

How the nervous systems appeared during evolution to become such a crucial feature of animals remains one of the most fascinating unanswered questions in biology. From the origin of neurons to their diverse organizations into nerve nets, brains, and nerve cords, the evolutionary mechanisms that led to the incredible diversity of nervous systems we can observe in nature are still mostly unknown. Many important questions are still debated among researchers to uncover the mysteries of the evolution of the nervous system. This chapter provides a short overview of the ongoing debates and hypotheses regarding the origin of neurons in early-branching metazoans as well as in some unicellular organisms. We describe the current knowledge about the organization and the unsuspected complexity of nerve nets in Cnidaria and highlight some of the current hypotheses about the centralization of nervous system in Bilateria.

8.1   Basic Concepts of Evolution

Evolution as described by Charles Darwin in the nineteenth century is the result of random variations among individuals conferring different values of fitness in a given environment, therefore increasing or reducing the probability of transmitting these traits

J. Duruz · S. G. Sprecher (✉)
Department of Biology, University of Fribourg, Fribourg, Switzerland
e-mail: jules.duruz@unifr.ch; simon.sprecher@unifr.ch

© Springer Nature Switzerland AG 2023

B. Egger (ed.), *Neurogenetics*, Learning Materials in Biosciences,
https://doi.org/10.1007/978-3-031-07793-7_8

to their progeny. The later discoveries of the mechanisms of heredity, the discovery of DNA, genomes and the way genes are transmitted within populations still fit with the ground principles originally laid out by Darwin. Besides the speciation process that gave rise to the diversity of organisms we observe today, scientists have also been interested in studying the evolution of specific anatomical features and cell types and comparing them between species to better understand the origin of such features. The neurons and nervous systems have been an important point of focus in these studies because of the very complex behavioral and cognitive capabilities that they confer. Nervous systems are only present in animals, although some early-branching animal clades have successfully evolved without bona fide neurons. The fundamental unit of any nervous system is the neuron: by relying on a mixture of electrical and chemical signaling and by its ability to form synapses, neurons have the ability to form complex networks with various degrees of complexity.

The ability to quickly react to stimuli from the environment and from other organisms is likely to have significantly improved the probability of an organism to survive and reproduce, therefore explaining why a large proportion of animals have maintained a nervous system throughout evolution. However, some major clades, such as Porifera (sponges), do not have neurons. These animals are sessile animals and are exclusively filter feeders. Their lifestyle, therefore, does not rely on quick responses to external stimuli or on navigation in complex environments and they have successfully survived and evolved independently for over 600 million years without neurons.

8.1.1 Terminology and Representation of Evolutionary Relationships

The basic evolutionary relationships of different clades of organisms can be represented with cladograms, which are a type of tree that represent the branching patterns between different clades. Unlike other phylogenetic trees, the time and sequence variation are not represented on a cladogram. As depicted in Fig. 8.1, each branch is called a **clade**. Depending on the tree, clades can represent any level of taxonomic organization (kingdom, phylum, class, order, family, genus, species). AB form a **monophyletic group**, which means that the two clades that form this group have one single common ancestor. E is the **last common ancestor** of A–B and C. C is the **sister group** of AB, which means that AB and C share one common ancestor. The order of all clades is interchangeable, as long as the branching points remain the same. A, B, C, and D are all currently living taxa and have all evolved for the same amount of time.

8.1.2 Evo-Devo: A Set of Comparative Methods to Uncover Characteristics of Common Ancestors

Evolutionary developmental biology (evo-devo) is a discipline of biology that compares the molecular development mechanisms across different species in order to gain informa-

Fig. 8.1 Example of a cladogram showing the evolutionary relationships of four clades (A, B, C, D) and E, the common ancestor of the clades A, B, and C

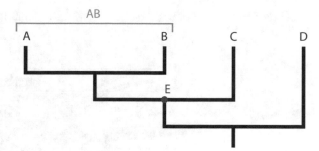

tion about their last common ancestor. These methods are based on the assumption that embryonic development and the way in which specific genes control these processes are very unlikely to have evolved several times in the exact same manner. Therefore, we can assume that a developmental process that relies on the same combination of genes between one animal clade and another is likely to have been present in their common ancestor, at the branching point of these clades. A typical example of this is represented by the conserved use of the transcription factor *PAX6*, in eye development across very distant animal clades (see Chap. 6) which indicates that the mechanisms of eye development are probably to a large degree shared between different animal clades, even though eyes have evolved several times independently.

8.2   Origin of Neurons and Synapses

Neurons and nervous systems are a very important feature that is highly specific to the Metazoa. Although Porifera and Placozoa do not have bona fide neurons, most animal phyla have nervous systems with an incredible variety of morphologies and organizations. While neurons are partly defined by their ability to propagate electrical signals, the components of synapses and the chemical signaling aspect of neurons seem to be better conserved across animals. Indeed, the conservation of core synaptic proteins among all Metazoa as well as some unicellular eukaryotes is very striking and has pushed evolutionary biologists to search for the origin of neurons in many early-branching clades of animals that are devoid of any visible nervous system.

8.2.1   The Evolution of Neurosecretory Proteins Predates the Emergence of Animals

The characterization of neurons has been predominantly based on morphological features such as the presence of a cell body, axons, dendrites, and synapses as well as their electrophysiological properties. In light of this type of characterization, animals were split into two groups: those with neurons and those without, with the underlying assumption that

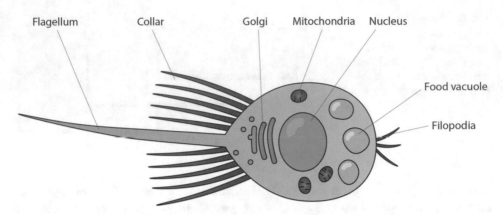

Flagellum   Collar   Golgi   Mitochondria   Nucleus

Food vacuole

Filopodia

Fig. 8.2 Anatomy of a choanoflagellate and its main organelles. The flagellum is used both for locomotion and feeding behavior

the earliest branching metazoans are the one without neurons and that neurons appeared once and have been maintained in all animal clades. However, advances in genomics, transcriptomics, and proteomics have provided important additional information and revealed that animals without neurons seem to possess some of the core molecular features that are generally associated with neurons. Furthermore, it was discovered that choanoflagellates—a group of unicellular protists—express some core components of synapses such as the SNARE machinery necessary for exocytosis of neurotransmitters; the choanoflagellate *Monosiga brevicollis* was shown to express the major components of the neurosecretory machinery SNAP-25, Syntaxin, Synaptobrevin, and Munc-18 [1]. This shows that many components of nervous systems and synapses actually predate the emergence of neurons even before the evolution of multicellular animals. The exact function of these proteins in unicellular organisms remains unknown. Additionally, recent discoveries describing light-induced synchronous behavioral response in colonies of choanoflagellates reinforce the possibility that they are capable of rapid cell-cell communications, similarly to a nervous system and could provide crucial information regarding the origin of multicellular organisms [2] (Fig. 8.2).

8.2.2 Placozoans Are Animals Without Neurons but with Cells That Share Homologies with Synapses

While nervous systems are a feature only present in animals, two major phyla do not possess any neurons and have therefore long thought to have diverged early from the rest of animals. The phylum Placozoa is particularly enigmatic due to its extreme morphological simplicity. The first described species of this phylum *Trichoplax adhaerens* have been shown to possess only six morphologically distinct cell types [3] (Fig. 8.3) and are only composed of two single layers of epithelia separated by a loose network

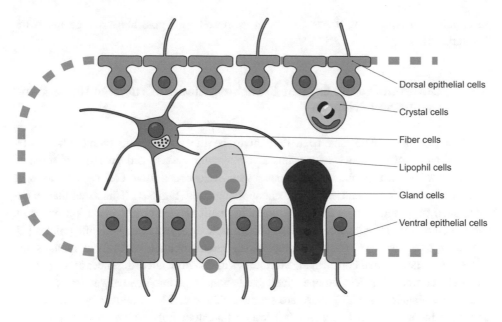

Fig. 8.3 The six described cell types of the placozoan *Trichoplax adhaerens* (adapted from Smith et al. [3]). The animal is composed of a dorsal epithelium and a ventral epithelium that comprise lipophil cells and gland cells. The middle part of the animal is composed of a loose network of fiber cells and few crystal cells

of fiber cells. These animals lack nervous systems, muscle and any identifiable organ systems nevertheless display a broad range of complex behaviors [4], proving that animal behavior and locomotion do not rely exclusively on the presence of a nervous system. The morphological simplicity of these animals pushed zoologists to think that placozoans might be a sister group to all other animals, making it the earliest branching metazoan. The sequencing and assembly of the *Trichoplax adhaerens* genome revealed the encoding of synaptic core proteins and further expression studies showed that their expression in placozoans is specific to one cell type: the gland cells [3, 5]. This suggested that either Placozoa have undergone secondary simplification of their body plan during evolution, which includes a loss of neurons or, alternatively, possess cells that derive from the ancestor of synapses. Placozoans in laboratory conditions only reproduce asexually and reports of embryonic development is anecdotic and poorly documented. Information about the embryonic development of placozoans could contribute to a better understanding of the origin of these strange animals. Sequencing of the genome of *Trichoplax adhaerens* however revealed the presence of many regulatory factors involved in animal development. However, without assessing the role of these factors in development, it is difficult to determine whether these transcription factors assumed a similar function in their last common ancestor. The phylogenetic position of Placozoan in the metazoan tree is debated and has been proposed as a sister group to all other animals due to its morphological simplicity. It has also been proposed as a possible sister group to the Cnidaria [6].

8.2.3   Determining the Earliest-Branching Clade Is Crucial to Understand Evolution of Neurons

Porifera (sponges) have long been thought to be the sister group to all other metazoans because of their lack of neurons and muscular system and because of some of the morphological resemblance of choanocytes—one constitutive cell type of sponges involved in filtration of nutrients—and choanoflagellates (Fig. 8.4). This resemblance has later been proposed to be a case of a possible result of convergent evolution based on the structural and functional differences between choanocytes and choanoflagellates [7]. However, extensive ultrastructural comparisons between choanocytes from sponges and choanoflagellates reveal both striking similarities and significant differences at the level of organelle composition, leaving open the debate about the homology between choanocytes and choanoflagellates [8]. Sponges are sessile animals and are exclusively filter-feeders and have therefore a relatively simple lifestyle that does not involve locomotion or the search for food.

Ctenophora (comb jellies), on the other hand, were thought to be closely related to Cnidaria because of their similar appearance and because of their nervous systems are composed of a diffuse network of neurons. The increasing amount of genomic data of

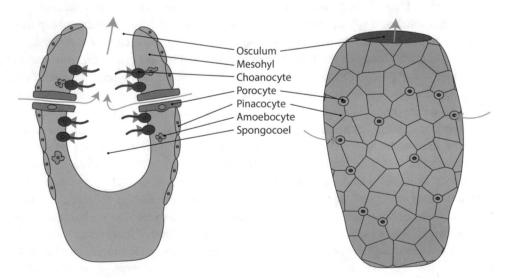

Fig. 8.4 Anatomy of a sponge (Porifera). Water passively enters through the pores formed by porocytes and nutrients are absorbed by choanocytes. Water flow (blue arrows) comes out through the osculum. Neurons are not present in sponges at all

both taxa has brought important questions regarding the initial assumption regarding the phylogenetic positions of Ctenophora and Porifera. It has been now proposed by many researchers that Ctenophora may actually be a sister group too all other metazoan [9–11]. The implication of this hypothesis would be that either nervous systems emerged independently in Ctenophora and Cnidaria in a case of convergent evolution of neurons or, alternatively, that the common ancestor of all animals had neurons that have then been lost in Placozoa and Porifera. Alternatively, recent studies also put Porifera as a sister group to all animals [12–14]. With this question still unresolved, one has to consider that different scenarios are possible for the evolutionary relationships between Porifera, Ctenophora, Placozoa, and Cnidaria (Fig. 8.5). Many more scenarios could be represented when considering that Placozoa are also considered a plausible sister group to other animals. However, most recent studies have either supported a Porifera-first scenario or a Ctenophora-first scenario.

8.3 The Nerve Nets of Cnidarians

Neurons can be organized in a very broad variety of configurations, but a distinction is generally made between centralized nervous systems that comprise a brain and one or several nerve cords, and nerve nets, in which neurons are organized diffusely across the body of the animal without any clear centralized structures. Cnidarians, a phylum that includes jellyfish, anemones, and corals, all possess nerve nets. These animals, many of which are active predators, display very complex behaviors that involve the processing of multisensory information from their environment, decision making, and even sleep-like behaviors such as described in the jellyfish *Cassiopea* [15]. This shows that centralization of the nervous system is not required to achieve these complex behaviors and therefore analyzing the link between neuronal networks within nerve nets and their link to specific behaviors is crucial to understanding how all nervous systems function across different animal clades.

Few cnidarian models have been used as laboratory biological models to study nerve nets. *Hydra vulgaris*, a freshwater polyp, has been studied for over 200 years for its astonishing regenerative capability. However, *Hydra* reproduces mainly asexually in laboratory conditions, which makes it very convenient to maintain genetically homogenous cultures but makes it less adapted to answering questions related mechanisms of embryonic development. The anemone *Nematostella vectensis* (Fig. 8.6) has been used in the laboratory to study gene expression and function during development [16]. *Nematostella* reproduces sexually in a way that can be directly induced by light and temperature in laboratory conditions [17], giving researchers access to large numbers of embryos to study development. Several genetic tools have been implemented in *Nematostella vectensis* to interfere with gene with RNA interference [18] and to genome editing with the CRISPR/Cas9 system [19] and the meganuclease system [20].

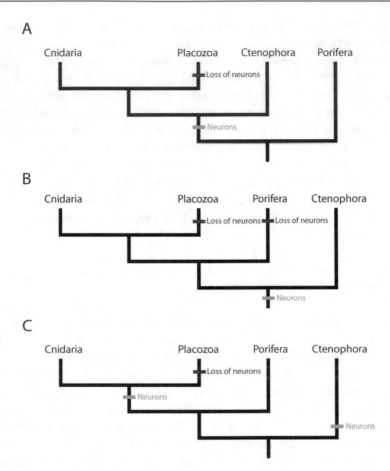

Fig. 8.5 Cladograms representing evolutionary scenarios: (**a**) Porifera is a sister group to all other animals. In that scenario, neurons emerge only once and are lost in Placozoa. (**b**) Ctenophora are a sister group to all other animals. In this scenario, neurons were present in the common ancestor of all animals and were lost two separate times in Porifera and Placozoa or (**c**) Ctenophora are a sister group to all other animals. In this scenario, neurons appear two separate times in evolution in a case of convergent evolution

Sequencing and analysis of the genome of *Nematostella vectensis* have revealed that these animals possess a lot of transcription factors that were initially thought to be typically involved in the formation of brains and nerve cords [21]. These findings spiked a renewed interest in the study of nerve nets with the knowledge that cnidarians possess many genes encoding for proteins involved in the formation of central nervous systems in bilaterian animals. The implication of that is that these genes were likely to be present in the cnidarian-bilaterian ancestor and it is crucial to understand whether these genes also play a role in the formation of nerve nets in order to infer on what nervous system of bilaterians and cnidarians may have been like.

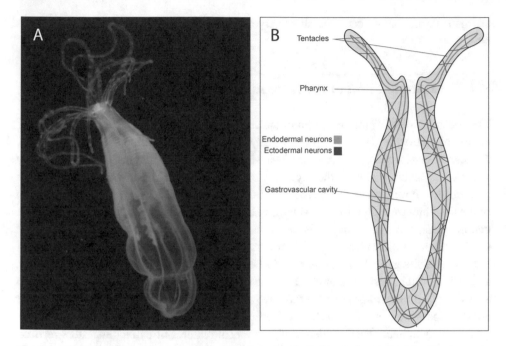

Fig. 8.6 Photograph of an adult Nematostella vectensis (left). Representation of the Nematostella nervous system with endodermal (green) and ectodermal (red) nerve nets

   The possibility to generate transgenic lines in both *Hydra* and *Nematostella* has enabled some important discoveries. A reporter line for LWamide (neuropeptide) neurons in *Nematostella* has revealed that this specific population of neurons develops in a stereotypical way, which means that the nerve net of this organism is not randomly organized but the same neurons always connect in a similar way during development [22]. This reinforces the idea that specific neural networks in Cnidaria serve specific functions that have biological relevance. Study of neuronal activity using a neuron-specific GCaMP transgenic line in *Hydra vulgaris* has also shown that distinct non-overlapping neuronal networks are involved in the responses to specific stimuli, indicating that specific parts of the nerve net are used to serve precise functions [23]. This is a characteristic that was only associated with centralized nervous systems in which an anatomical part corresponds to a specific function, but it was shown that this also applied in the nerve nets of cnidarians. Technological advances that enable the sequencing of transcriptomes at a single-cell resolution enabled the study of individual cnidarian cell types. This gave researchers the opportunity to look into the neuronal diversity of entire animals and associate these neuronal types with specific gene expression. These studies revealed, using both *Hydra* and *Nematostella* that cnidarians have an unsuspected neuronal diversity [24, 25], indicating that many types of neurons could be serving different purposes in the nerve nets of cnidarians, breaking with the idea of a "simple" nerve net and suggesting that there are many things that we can learn about nervous systems in general from the study of nerve nets.

8.4   Centralization of the Nervous System in Bilaterians

The recent studies of non-bilaterian organisms tend to indicate that the level of centralization of a nervous system if not necessarily linked to genetic complexity (genome size, number of genes, etc.). In fact, it has been shown that non-bilaterians have evolved very successfully and are able to display complex behaviors without having central nervous systems. Moreover, genes that were proposed to be specifically involved in the formation of the CNS and were therefore considered to be bilaterian novelties have been found in the genomes of a variety of non-bilaterian animals. This makes the study of the origin of the CNS particularly difficult because the presence of CNS-linked genes in different species of interest does not indicate the presence of a CNS in their common ancestor. Consequently, a very important question remains about the evolution of the CNS: Did the last common ancestor of all bilaterians (Urbilateria) have a centralized nervous system and a brain, or did the process of centralization happen several times during evolution?

For many animal clades evidence of anatomical features of extinct ancestors rests on the analysis of fossil records. In the case of animals only composed of soft tissues and in the absence of mineralized structures (bones, shells, etc.) fossils are very unlikely to be found and consist only of traces left on substrates. The oldest uncontested bilaterian fossil is *Kimberella* sp., which was dated at 555–558 million years ago [26]. Other possible older bilaterian fossils have been found but the taxonomy could not be confidently established. It is unclear whether these fossils are indeed bilaterians and sometimes even if they are animals. Because of the difficulty to obtain reliable fossil record, we are almost exclusively reliant on other sources of information to gain insights into the Urbilaterian features.

Generally, two opposed scenarios are debated: either the Urbilateria was an animal with various key anatomical features found in many bilaterians (Segmentation, CNS, through-gut) that have been maintained throughout the bilaterian clades (Fig. 8.7a). This would imply that these features shared in so many clades are all diverged from a single ancestor, making them homologous.

On the other hand, the urbilaterian could have been an animal with simpler morphology, closer to the anatomy of non-bilaterians (Nerve net, single mouth opening, no organ systems). This would imply that typically bilaterian features such as CNS, through gut and segmentation could have appeared several times during evolution, following the split of Deuterostomia and Protostomia (Fig. 8.7b). Alternatively, to these two opposed views, different combinations of these features have been proposed.

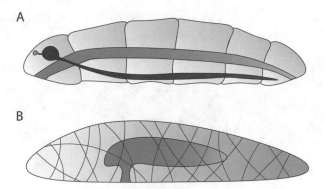

Fig. 8.7 Two scenarios for the anatomy of the urbilaterian. The gut is shown in green and the nervous system in red. (**a**) Urbilateria with a segmented body, a through-gut, a brain with a sensory organ (eye) and a ventral nerve cord. (**b**) Urbilateria with a single gut opening and a diffuse nerve net

8.4.1 The Study of Expression Domains Across Species Can Provide Information About CNS Origins

Given the lack of fossil data for nervous system in general, the study of urbilaterian characteristics relies heavily on the search for homologies in gene expression domains between different species of animals. Some transcription factors have shown homologous expression domains in Cnidaria and Bilateria. One example was presented in [27] (Fig. 8.8) where striking similarities in the expression of some selected markers is similar between late gastrula of frogs, the larva of the annelid *Platynereis* and the planula larva of *Nematostella*. They highlight similar expression of the sensory markers Sine Oculis homeobox 3 (SIX3) and Retinal homeobox (Rx) in the aboral region of the *Platynereis* and *Nematostella* larvae and the animal pole of the *Xenopus* gastrula indicating possible homology between the head region and eyes of bilaterians and the aboral region of cnidarians. On the opposite side of this "sensory" region similar expression patterns of the transcription factors *sonic hedgehog* (HH) and FoxA, which trigger the differentiation of monoaminergic neurons in vertebrates, are observed together with *brachyury*, known for its role in the blastopore formation. Such comparisons of expression domains of orthologous genes constitute some of the most valuable and widespread tools for evo-devo research. The use of In Situ hybridizations, a highly replicable method that consists of hybridizing RNA strands to detect gene expression is a very efficient way to obtain this data in a way that does not require the use of genetic manipulations and can therefore be applied on many different taxa for broader comparisons.

A hypothesis proposed for the origin of bilateral symmetry in animals is that the formation of gastric pouches to increase the surface contact with nutrients on the digestive epithelium [28] (Fig. 8.9). In typical non-bilaterian animals, the digestive system consists of a single large cavity in which food is digested. It was suggested that the digestive cavities

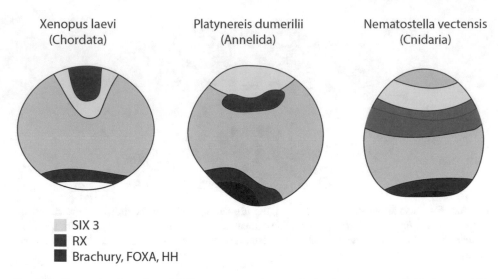

| Xenopus laevi | Platynereis dumerilii | Nematostella vectensis |
| (Chordata) | (Annelida) | (Cnidaria) |

SIX 3
RX
Brachury, FOXA, HH

Fig. 8.8 Expression domains of different groups of transcription factors across Cnidaria, Annelids (Protostomia), and Vertebrates (Deuterostomia) (adapted from Arendt et al. [27])

have become narrower to increase the surface available to absorb nutrients with gastric pouches controlling the water flow, breaking the radial symmetry of the animal. Closure of this slit-like mouth opening in the middle would then leave a mouth and anus and would favor a one-way water flow. This hypothesis would suggest that bilateral symmetry in animals arose before the split of Cnidaria and Bilateria [29].

The question of centralization of the nervous system has been very important in the study of nervous system evolution. While many bilaterian nervous systems could be characterized as centralized, they show a striking diversity of organizations. Some patterns can be observed within different groups and despite numerous exceptions, Protostomia generally have a ventral nerve cord while Deuterostomia have a dorsal nerve cord. However, hemichordates, which are deuterostomes, have a very interesting nervous system: they possess both a ventral and a dorsal nerve cord as well as a diffuse nerve net. From these patterns of nervous system organizations, different hypotheses can be made regarding the organization of the nervous system of the urbilaterian [30] (Fig. 8.10). These hypotheses suggest that the Urbilaterian could either have had several nerve chords, one of which is homologous to the vertebrate nerve cord and another that is homologous to the protostome nerve cord. Alternatively, the urbilaterian could have had either a unique ventral or a dorsal nerve chord and inversions could have happened throughout evolution. Lastly, it is possible that the urbilaterian did not have a centralized nervous system and that the process of centralization happened after the split of deuterostomes and protostomes, therefore explaining the differences in their organization.

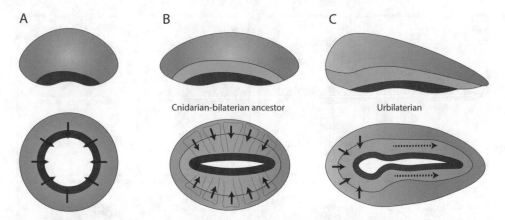

Fig. 8.9 Schematic representations of the steps of evolution according to the hypothesis that bilateral symmetry originated from the formation of gastric pouches (adapted from Arendt et al. [28]). (**a**) In the neuralian ancestor with a ring of neurons (red) around the mouth opening that control feeding. (**b**) The cnidarian-bilaterian ancestor has gastric pouches that control the water flow with neurons from the sensory-integrative center (yellow) toward a slit-shaped mouth opening. (**c**) Urbilaterian with the motor center of the nerve cord (red) and sensory-integrative center (yellow). Arrows indicate the direction of the water flow

8.4.2   Xenacoelomorpha Are Worms That Resemble the Urbilateria

Xenacoelomorpha are a phylum composed of Xenoturbellidae and Acoelomorpha [31]. They are marine worms that display a broad range of morphological diversity. Their anatomy is relatively simple and lacks a coelom and they have a single gut opening. The morphology of Xenacoelomorphs is raising particular interest among evolutionary biologists because it is consistent with hypotheses regarding the anatomy of the urbilaterian. Indeed, the Xenacoelomorpha possess anatomical features that resemble non-bilaterian features such as a single mouth opening, lack of segmentation and appendages and the sole use of cilia for locomotion. On the other hand, they possess features that are typical Bilateria such as centralized nervous systems. This phylum has been proposed as a sister group to all other bilaterian. In this case, the last common ancestor of Xenacoelomorpha and other bilaterians would be the Urbilateria and it would therefore make Xenacoelomorpha particularly interesting to study the origin of the CNS in bilaterians. This hypothesis was contested with further phylogenetic analyses to propose that Xenaceolomorpha belongs to the Deuterostomia [31]. Additional genetic data with increased sampling later revealed that Xenacoelomorpha are more likely to be, as initially thought, a sister group to other bilateria [32]. This was later again determined to be a result of systematic error in the construction of phylogenetic tree and corrections taking into account the particularly fast evolutionary rate of some genes determined that Xenacoelomorpha would be Deuterostomes again [33]. This ongoing debate is of particular relevance for the study of the evolution and origin of the CNS. If Xenacoelomorpha are a sister group to all other bilaterians then the presence

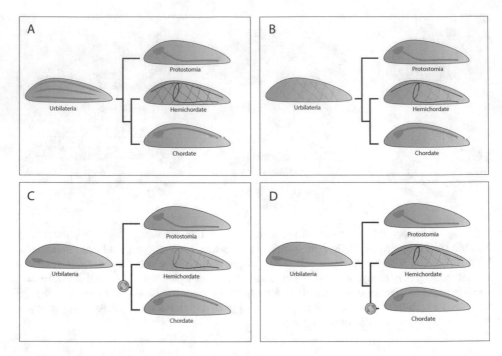

Fig. 8.10 Proposed scenarios of the evolution of nervous system organization in Urbilateria, Protostomia, Hemichordata, and Chordata. Circular arrows indicate dorsal/ventral inversion of the nerve cord (adapted from Holland [30]). (**a**) Urbilateria with several nerve cords. The urbilaterian ventral cord is homologous to the nerve cord of protostomes and the urbilaterian dorsal chord is homologous to the nerve cord of chordates. In that scenario the homologies in the hemichordate nervous system is unclear. (**b**) Urbilaterian nerve net with homology to the hemichordate nerve net. (**c**) Urbilateria with a ventral nerve cord that is homologous to the nerve cord of other bilaterians. In this, scenario there is a ventral/dorsal inversion at the base of deuterostomes which implies that the dorsal cord of hemichordates is homologous to the one from chordates. (**d**) Urbilateria with a ventral nerve cord that is homologous to the nerve cord of other bilaterians. In this, scenario there is a ventral/dorsal inversion at the base of chordates, which implies that the ventral cord of hemichordates is homologous to the one from chordates

of a CNS in many different species could be an indication that the CNS is a feature that was already present in the Urbilateria. Alternatively, it could mean that CNS have evolved several times during evolution. Independently of their phylogenetic position, it has been proposed that the central nervous system of acoels may be the result of independent centralization [34, 35]. The nervous systems of some Xenacoelomorpha have been studied and described mainly in the acoels *Isodiametra pulchra* [36], *Symsagittifera roscoffensis* [37–39], and *Hofstenia miamia* [40] (Fig. 8.11).

Having diverged very early from all other bilaterians, the study of Xenacoelomorpha can provide very valuable information to better understand the origin and evolution of bilaterian innovations such as the brain and nerve cords.

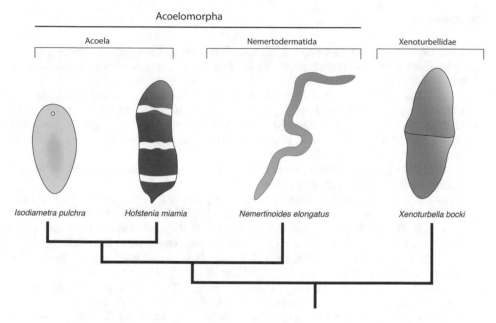

Fig. 8.11 Phylogeny of Xenacoelomorpha. The subphylum Acoelomorpha is subdivided into two classes Acoela and Nemertodermatida and the subphylum Xenoturbellidae has only one represented genus Xenoturbella

Take-Home Message

The study of the origin of neurons and the process that led neurons to organize into nerve nets and central nervous system in animals remains a widely open and fascinating debate. The pursuit to answer these fundamental questions about the origins of nervous systems could provide us with a better understanding not only of the process of evolution as a whole but also about how the nervous system works in its different forms. Many questions in this area of research remain debated and unsolved and much more research, sampling more animals is necessary to get closer to reconstructing the full picture of nervous system evolution. Nevertheless, the studies of genomes, transcriptomes, gene expression patterns, and gene function across many clades of animals, enhanced by major technological advances, have enabled the discovery of unsuspected homologies between various clades of animals and have significantly reduced the number of possible scenarios that depict the main steps of nervous system evolution.

References

1. Burkhardt P, Stegmann CM, Cooper B, Kloepper TH, Imig C, Varoqueaux F, et al. Primordial neurosecretory apparatus identified in the choanoflagellate *Monosiga brevicollis*. Proc Natl Acad Sci U S A. 2011;108(37):15264–9.
2. Brunet T, Larson BT, Linden TA, Vermeij MJA, McDonald K, King N. Light-regulated collective contractility in a multicellular choanoflagellate. Science. 2019;366(6463):326–34.
3. Smith CL, Varoqueaux F, Kittelmann M, Azzam RN, Cooper B, Winters CA, et al. Novel cell types, neurosecretory cells, and body plan of the early-diverging metazoan Trichoplax adhaerens. Curr Biol. 2014;24(14):1565–72.
4. Heyland A, Croll R, Goodall S, Kranyak J, Wyeth R. *Trichoplax adhaerens*, an enigmatic basal metazoan with potential. Methods Mol Biol. 2014;1128:45–61.
5. Srivastava M, Begovic E, Chapman J, Putnam NH, Hellsten U, Kawashima T, et al. The Trichoplax genome and the nature of placozoans. Nature. 2008;454(7207):955–60.
6. Laumer CE, Gruber-Vodicka H, Hadfield MG, Pearse VB, Riesgo A, Marioni JC, et al. Support for a clade of Placozoa and Cnidaria in genes with minimal compositional bias. Elife. 2018;7
7. Mah JL, Christensen-Dalsgaard KK, Leys SP. Choanoflagellate and choanocyte collar-flagellar systems and the assumption of homology. Evol Dev. 2014;16(1):25–37.
8. Laundon D, Larson BT, McDonald K, King N, Burkhardt P. The architecture of cell differentiation in choanoflagellates and sponge choanocytes. PLoS Biol. 2019;17(4):e3000226.
9. Ryan JF, Pang K, Schnitzler CE, Nguyen AD, Moreland RT, Simmons DK, et al. The genome of the ctenophore Mnemiopsis leidyi and its implications for cell type evolution. Science. 2013;342(6164):1242592.
10. Moroz LL, Kocot KM, Citarella MR, Dosung S, Norekian TP, Povolotskaya IS, et al. The ctenophore genome and the evolutionary origins of neural systems. Nature. 2014;510(7503):109–14.
11. Whelan NV, Kocot KM, Moroz TP, Mukherjee K, Williams P, Paulay G, et al. Ctenophore relationships and their placement as the sister group to all other animals. Nat Ecol Evol. 2017;1(11):1737–46.
12. Feuda R, Dohrmann M, Pett W, Philippe H, Rota-Stabelli O, Lartillot N, et al. Improved modeling of compositional heterogeneity supports sponges as sister to all other animals. Curr Biol. 2017;27(24):3864–70 e4.
13. Pisani D, Pett W, Dohrmann M, Feuda R, Rota-Stabelli O, Philippe H, et al. Genomic data do not support comb jellies as the sister group to all other animals. Proc Natl Acad Sci U S A. 2015;112(50):15402–7.
14. Simion P, Philippe H, Baurain D, Jager M, Richter DJ, Di Franco A, et al. A large and consistent phylogenomic dataset supports sponges as the sister group to all other animals. Curr Biol. 2017;27(7):958–67.
15. Nath RD, Bedbrook CN, Abrams MJ, Basinger T, Bois JS, Prober DA, et al. The jellyfish cassiopea exhibits a sleep-like state. Curr Biol. 2017;27(19):2984–90 e3.
16. Layden MJ, Rentzsch F, Rottinger E. The rise of the starlet sea anemone *Nematostella vectensis* as a model system to investigate development and regeneration. Wiley Interdiscip Rev Dev Biol. 2016;5(4):408–28.
17. Stefanik DJ, Friedman LE, Finnerty JR. Collecting, rearing, spawning and inducing regeneration of the starlet sea anemone, *Nematostella vectensis*. Nat Protoc. 2013;8(5):916–23.
18. Calcino AD, Fernandez-Valverde SL, Taft RJ, Degnan BM. Diverse RNA interference strategies in early-branching metazoans. BMC Evol Biol. 2018;18(1):160.
19. Ikmi A, McKinney SA, Delventhal KM, Gibson MC. TALEN and CRISPR/Cas9-mediated genome editing in the early-branching metazoan *Nematostella vectensis*. Nat Commun. 2014;5:5486.

20. Renfer E, Technau U. Meganuclease-assisted generation of stable transgenics in the sea anemone *Nematostella vectensis*. Nat Protoc. 2017;12(9):1844–54.
21. Putnam NH, Srivastava M, Hellsten U, Dirks B, Chapman J, Salamov A, et al. Sea anemone genome reveals ancestral eumetazoan gene repertoire and genomic organization. Science. 2007;317(5834):86–94.
22. Havrilak JA, Faltine-Gonzalez D, Wen Y, Fodera D, Simpson AC, Magie CR, et al. Characterization of NvLWamide-like neurons reveals stereotypy in *Nematostella* nerve net development. Dev Biol. 2017;431(2):336–46.
23. Dupre C, Yuste R. Non-overlapping neural networks in *Hydra vulgaris*. Curr Biol. 2017;27(8):1085–97.
24. Sebe-Pedros A, Saudemont B, Chomsky E, Plessier F, Mailhe MP, Renno J, et al. Cnidarian cell type diversity and regulation revealed by whole-organism single-cell RNA-seq. Cell. 2018;173(6):1520–34 e20.
25. Siebert S, Farrell JA, Cazet JF, Abeykoon Y, Primack AS, Schnitzler CE, et al. Stem cell differentiation trajectories in Hydra resolved at single-cell resolution. Science. 2019;365(6451)
26. Fedonkin MA, Waggoner BM. The Late Precambrian fossil Kimberella is a mollusc-like bilaterian organism. Nature. 1997;388(6645):868–71.
27. Arendt D, Tosches MA, Marlow H. From nerve net to nerve ring, nerve cord and brain—evolution of the nervous system. Nat Rev Neurosci. 2016;17(1):61–72.
28. Arendt D, Benito-Gutierrez E, Brunet T, Marlow H. Gastric pouches and the mucociliary sole: setting the stage for nervous system evolution. Philos Trans R Soc Lond B Biol Sci. 2015;370(1684)
29. Finnerty JR. Did internal transport, rather than directed locomotion, favor the evolution of bilateral symmetry in animals? Bioessays. 2005;27(11):1174–80.
30. Holland LZ. Evolution of basal deuterostome nervous systems. J Exp Biol. 2015;218(Pt 4):637–45.
31. Philippe H, Brinkmann H, Copley RR, Moroz LL, Nakano H, Poustka AJ, et al. Acoelomorph flatworms are deuterostomes related to *Xenoturbella*. Nature. 2011;470(7333):255–8.
32. Cannon JT, Vellutini BC, Smith J 3rd, Ronquist F, Jondelius U, Hejnol A. *Xenacoelomorpha* is the sister group to *Nephrozoa*. Nature. 2016;530(7588):89–93.
33. Philippe H, Poustka AJ, Chiodin M, Hoff KJ, Dessimoz C, Tomiczek B, et al. Mitigating anticipated effects of systematic errors supports sister-group relationship between *Xenacoelomorpha* and *Ambulacraria*. Curr Biol. 2019;29(11):1818–26 e6.
34. Gavilan B, Perea-Atienza E, Martinez P. *Xenacoelomorpha*: a case of independent nervous system centralization? Philos Trans R Soc Lond B Biol Sci. 2016;371(1685):20150039.
35. Martin-Duran JM, Pang K, Borve A, Le HS, Furu A, Cannon JT, et al. Convergent evolution of bilaterian nerve cords. Nature. 2018;553(7686):45–50.
36. Achatz JG, Martinez P. The nervous system of Isodiametra pulchra (Acoela) with a discussion on the neuroanatomy of the *Xenacoelomorpha* and its evolutionary implications. Front Zool. 2012;9(1):27.
37. Bery A, Cardona A, Martinez P, Hartenstein V. Structure of the central nervous system of a juvenile acoel, *Symsagittifera roscoffensis*. Dev Genes Evol. 2010;220(3-4):61–76.
38. Perea-Atienza E, Gavilan B, Chiodin M, Abril JF, Hoff KJ, Poustka AJ, et al. The nervous system of *Xenacoelomorpha*: a genomic perspective. J Exp Biol. 2015;218(Pt 4):618–28.
39. Semmler H, Chiodin M, Bailly X, Martinez P, Wanninger A. Steps towards a centralized nervous system in basal bilaterians: insights from neurogenesis of the acoel *Symsagittifera roscoffensis*. Dev Growth Differ. 2010;52(8):701–13.
40. Hulett RE, Potter D, Srivastava M. Neural architecture and regeneration in the acoel *Hofstenia miamia*. Proc Biol Sci. 1931;2020(287):20201198.

Embryonic Neurogenesis in the Mammalian Brain

Dotun Adeleye Adeyinka, Boris Egger

What You Will Learn in This Chapter

The mammalian brain is probably the most fascinating and complex organ that has evolved over millions of years. In this chapter, we learn about some key genetic factors that control mammalian brain development with a focus on the cerebral cortex. Selected topics highlight oscillatory expression of neurogenic and proneural factors, the relationship between cell cycle control and cell fate and the spatiotemporal generation of neurons in the layered cortex. In a second part, we introduce the neural stem and progenitor cell types in the mammalian neocortex that potentially are the key recent inventions to distinguish higher evolved gyrencephalic from more primitive lissencephalic brains. We discuss the concepts and cellular mechanisms that might have led to neocortex expansion during evolution toward the primate brain.

9.1 Molecular and Cellular Mechanisms of Neural Development

In vertebrates, including mammals, neurogenesis takes place in the neuroepithelium of the folded neural tube. Neuroepithelial cells proliferate in the so-called germinal or ventricular zone near the inner lumen (ventricle) to form the neural tube. Neuroepithelial cells are polarized within a single-layered cell sheet. The basal side of the neuroepithelial cells

D. A. Adeyinka · B. Egger (✉)
Department of Biology, University of Fribourg, Fribourg, Switzerland
e-mail: adeleyedotun.adeyinka@unifr.ch; boris.egger@unifr.ch

© Springer Nature Switzerland AG 2023
B. Egger (ed.), *Neurogenetics*, Learning Materials in Biosciences,
https://doi.org/10.1007/978-3-031-07793-7_9

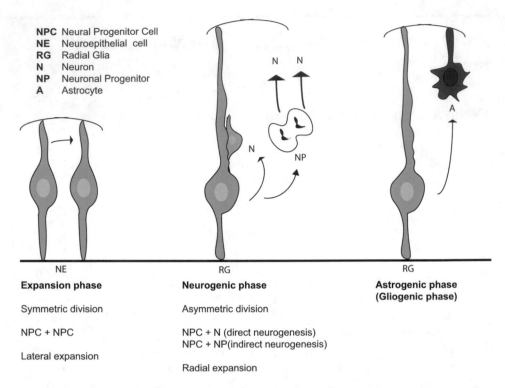

NPC Neural Progenitor Cell
NE   Neuroepithelial cell
RG   Radial Glia
N   Neuron
NP   Neuronal Progenitor
A   Astrocyte

NE   RG   RG

Expansion phase   **Neurogenic phase**   **Astrogenic phase**
                                                        (Gliogenic phase)

Symmetric division   Asymmetric division

NPC + NPC   NPC + N (direct neurogenesis)
                           NPC + NP(indirect neurogenesis)

Lateral expansion

                           Radial expansion

Fig. 9.1 Progression from expansion phase to neurogenesis phase. Neuroepithelial cells expand the pool of neural stem cells through symmetric proliferative divisions. Radial glial cells undergo self-renewing asymmetric divisions generating neurons either directly or via an intermediate progenitor cell. Following the neurogenesis radial glial cells serve as progenitors during a gliogenic phase to generate astrocytes. Redrawn from [3]

contacts the basal lamina and the apical end of each cell abuts the ventricle of the neural tube. Symmetric cell division initially leads to the expansion of neural epithelium and the neural progenitor pool. Prior to the onset of neurogenesis, neuroepithelial cells are transformed into radial glial cells and start to divide in a stem cell-like asymmetric mode to self-renew and generate differentiating neurons directly, as well as additional intermediate progenitor cell types (Fig. 9.1) [1, 2].

A special characteristic of neuroepithelial and radial glial cells is a process that is called interkinetic nuclear migration. In this process, the cell nucleus moves up and down within the cytoplasm of the elongated and narrow cell body, whereby the position of the nucleus varies in relation to the phases of the mitotic cell cycle. In G1 of the cell cycle, cell nuclei occupy the middle of the ventricular zone. Nuclei move upwards to the top of these zones and duplicate their DNA in S-phase. As cells complete DNA replication and enter G2, the nuclei descend to the ventricular surface, where they proceed through mitosis [4]. Since radial glial cells have also supporting properties for migrating neurons and express glia-specific markers they were considered for a long time as part of the glial lineage. However, radial glial cells share a number of common features with cycling progenitors, including interkinetic nuclear migration, expression of a neural precursor marker and a characteristic radial morphology [5]. It became clear that radial glial cells are the major source of both neuronal and glial progenitors in the ventricular zone [6]. Furthermore, expression data suggest that different subpopulations of radial glial cells exist, which have neurogenic, gliogenic, or neurogliogenic features [7]. Later in this chapter, we will discuss different subpopulations of radial glial cells, which are found in the developing mammalian cortex, including primates.

9.1.1 Symmetric Versus Asymmetric Neural Stem Cell Division Modes

Neuroepithelial cells and radial glial cells employ similar mechanisms as *Drosophila* neuroepithelial cells and neuroblasts to regulate cell division modes. Spindle orientation and the segregation of cell fate determinants are crucial to control daughter cell fates. Initially, neuroepithelial cells divide with a vertical cleavage plane (horizontal spindle axis) to proliferate and generate equal daughter cells. Thereby, the basal elongated process, also called radial glial fiber, is split and divided among both daughter cells or inherited by one daughter cell while the other daughter re-extending it de novo. Mammalian homologs of the Par3/Par6/aPKC localize to the apical cortex and are symmetrically distributed between daughter cells. Notch activity is also maintained in both daughter cells and maintains the neuroepithelial cell state. Neuroepithelial cells transition to radial glial cells. At the onset of neurogenesis radial glial cells divide asymmetrically whereby the basal process is inherited by the self-renewed mother cell only, which retains the undifferentiated state. The differentiating daughter cell loses its contact at the apical ventricle and adopts a neuronal fate either directly or indirectly over an intermediate progenitor cell. Unlike *Drosophila* neuroblast, radial glial cells display not a horizontal but rather an oblique cleavage plane during asymmetric cell division. Nevertheless, it leads to the asymmetric distribution of cell fate determinants ie. Par3/Par6/aPKC and to the unequal activation of Notch signaling. In the self-renewed mother cell, Notch activity tends to be maintained high while in the differentiating daughter cell (neurons or intermediate progenitors) Notch signaling is downregulated. The newborn neurogenic daughter cells withdraw their apical end feet from the apical side and move towards the subventricular zone [8, 9].

Disruption of factors that are controlling spindle orientation has drastic consequences for neurogenesis. Misorientation of the mitotic spindle might result in the depletion of neural progenitor cells. For example, LGN, which shares sequence identity with *Drosophila* Pins promotes planar spindle orientations in the neocortex and loss-of-function leads to random spindle orientations. In this case, the number of apical radial glial cells is decreased while the number of intermediate progenitors is increased. Interestingly, disruption of LGN function also leads to an increase of radial glial cells that are displaced towards the subventricular zone and resemble the population of basal or outer radial glial cells. mInsc, the mouse homolog of *Drosophila* Insc has also been implicated in the regulation of the mitotic spindle. It acts during later neurogenic phases to inhibit planar divisions and to promote oblique and vertical orientations for the production of intermediate progenitors and basal radial glial cells [10, 11].

Mutations in several proteins involved in the formation of centrosomes and mitotic spindles assembly and orientation have been implicated in human brain diseases. Notably, mutant alleles of MCPH1, ASPM, and CDK5RAP2 can lead to microcephaly, a condition characterized by abnormally small brains at birth [12, 13].

9.1.2   Proneural Genes in Vertebrate Nervous System Development

A number of bHLH transcription factors are involved in early neurogenesis. Several genes that are related to *Drosophila acheate (ac), scute (sc),* and *atonal (ato)* genes were identified in vertebrates, including mammalian species. The vertebrate *acheate-scute* (*asc*) family includes *Ash1*, which is present in all species studied to date, and three other genes that have each been found in only one vertebrate species [14, 15]. Several *ato*-related genes are present in the vertebrate genome but only two (*Math1* and *Math5* in the mouse) are considered to be orthologs of *Drosophila atonal.* Other vertebrate *ato*-related genes can be grouped into distinct families that share family specific amino acid residues in their bHLH domain. Three of these groups are the *Neurogenin* (*Ngn*) family, the *Neuro D* family, and the *Olig* family. Most vertebrate proneural-related genes are expressed primarily or exclusively in the developing nervous system. However, similarly to the situation in *Drosophila,* not all proneural-related genes have proneural function. Among the genes in the *asc, ato,* and *Ngn* families in the mouse, mutations in *Mash1, Ngn1,* and *Ngn2* show severe phenotypes during early neurogenesis in various brain regions. These impairments are associated with a loss of progenitor populations and a failure to express Delta and another Notch ligand, Serrate/Jagged [16–19]. Along these lines, overexpression of a Xenopus *Ngn*-related gene in the surface ectoderm of *Xenopus* embryos leads to extensive ectopic neurogenesis and is able to induce *Delta1* expression [19]. Thus, as in *Drosophila*, loss- and gain-of-function experiments of vertebrate proneural genes produce opposite phenotypes. Moreover, these experiments also demonstrate that proneural genes interact with the Delta/Notch signaling cascade in vertebrate neuroepithelial cells. Since not all CNS progenitors are affected by mutations in known proneural genes, this suggests that in vertebrates, as in *Drosophila*, additional genes with proneural function exist.

9.1.3   Notch Signalling in Vertebrate Nervous System Development

Genes that have crucial roles at each step of Notch signaling were found in various vertebrate species: *Notch, Delta, Su(H)/RBP-Jk,* and *E(spl)/Hes* [20]. A number of experiments in *Xenopus* and mouse demonstrate that the Notch signaling cascade is involved in cellular outputs that are similar to those described for *Drosophila.* For example, in the *Xenopus* tadpole, ectopic activation of Notch signaling, achieved by the expression of *Delta1* or the intracellular domain of *Notch1*, leads to a reduced number of primary neurons. In contrast, when signaling is inhibited by the expression of dominant negative forms of Notch signaling components (i.e., dominant negative Delta) primary neurogenesis is markedly enhanced [21–23]. Comparable observations were also made in murine CNS development. In the mouse, four Notch-like receptors and two Delta-like ligands have been identified. Inactivation of Notch signaling components has demonstrated a function for *Notch* in blocking neuronal differentiation. Mouse models with loss-of-function for *Notch1* and *Su(H)/RBP-Jk* show downregulation of the Enhancer of split family member *Hes5* (but not *Hes1*) and upregulation of proneural genes such as *Mash1*. Furthermore, in the absence of *Notch* function, upregulation for neuronal differentiation genes such as *NeuroD* was shown. These experiments support a model wherein Notch signaling prevents neural stem cells from entering neuronal differentiation [24]. Indeed, rather than discriminating between two alternative cell types, as seen in *Drosophila* embryonic neuroectoderm during lateral inhibition, the principal function of Notch signaling in the early mammalian CNS is thought to maintain cells in an undifferentiated state [25].

9.1.4 Oscillation of Hes and Proneural Factors

Hes and proneural bHLH factors show a dynamic expression pattern during neural proliferation and differentiation [26]. It leads to a so-called salt-and-pepper pattern of expression in the early cerebral cortex. Neurogenic Hes factors promote neural stem cell maintenance, while repressing proneural factor induced neuronal production and differentiation. Elegant luciferase reporter assays in vitro and in vivo revealed that in neural progenitor cells the levels of Hes1 and Ascl1 proteins oscillate within 2–3-h periods [27]. In contrast, neurogenesis coincides with the sustained expression of proneural factor Ascl1 and the repression of Hes1. A hypothesis here is that the periodic Ascl1 expression is sufficient to activate genes involved in cell cycle progression, but that sustained expression is required for genes involved in cell cycle exit and differentiation (Fig. 9.2). Ngn2 and Dll1 are oscillating and neural progenitors are at one time point signal receivers and at the next time point signal senders. The observations led to a revised model of lateral inhibition, whereby it is rather required for the maintenance of neural progenitor states instead as for neural selection [26].

9.1.5 Cross-Regulation Between the Cell Cycle and Cell Fate

During neurogenesis, there is an intrinsic relation between the establishment of progenitor cell states and cell cycle progression. It appears that one of the key determinants of neurogenesis is cell cycle length (Fig. 9.3). Comparative studies observing neurogenic and proliferative neural progenitors led to the cell cycle length hypothesis [29]. It

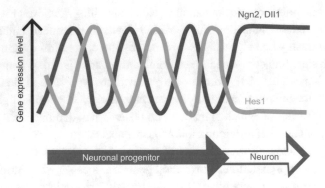

Fig. 9.2 Expression of *Hes1*, *Ngn2*, and *Dll1* oscillates in dividing neural progenitors. In immature postmitotic neurons, *Hes1* is downregulated, whereas *Ngn2* and *Dll1* are upregulated in a sustained manner. Oscillatory expression of Ngn2 seems not sufficient but sustained upregulation is required for neuronal differentiation. Redrawn from [28]

suggests that proliferative progenitors undergo rapid cell cycle progression and that lengthening of the G1 phase of the cell cycle is an instructive measure to initiate neuronal differentiation and the depletion of neural progenitors. Indeed, the forced shortening of G1 via Cdk4/CyclinD1 overexpression promotes the expansion of neural progenitors and leads to delayed neurogenesis [30].

9.1.6   Spatiotemporal Generation of Postmitotic Neurons

The outcome of proliferation, differentiation, and migration of neuronal cells in vertebrate neurogenesis is a highly organized layered structure. Studies in the developing cerebral cortex have revealed that, in principle, the cortical layers are generated in an inside-out manner of development [32]. The first neurons to be born constitute an exception; these neurons leave the ventricular zone to populate a single layer near the pial surface of the developing brain, called the preplate. These neurons are subsequently split into two layers by the formation of the cortical plate. The upper layer is called the marginal zone and will become layer 1 of the adult animal. The deeper layer constitutes the subplate and becomes located beneath the cortical plate. In an intermediate position between the ventricular zone and the subplate the mantle layer (or intermediate zone) and later the subventricular zones form where neuronal and glial progeny are transiently located and commence differentiation. The genesis of cortical plate layers 2–6 follows and is characterized by a highly organized procession of cell birth, migration, and differentiation. The cells of the deepest layer, layer 6, are the earliest to become postmitotic and migrate out, inserting themselves between the nascent marginal zone and subplate neurons. Then, the cells of the more superficial layers, layer 2–5, are generated [4]. The molecular mechanisms, which control the timing and the formation of the layered cerebral structure are actively studied.

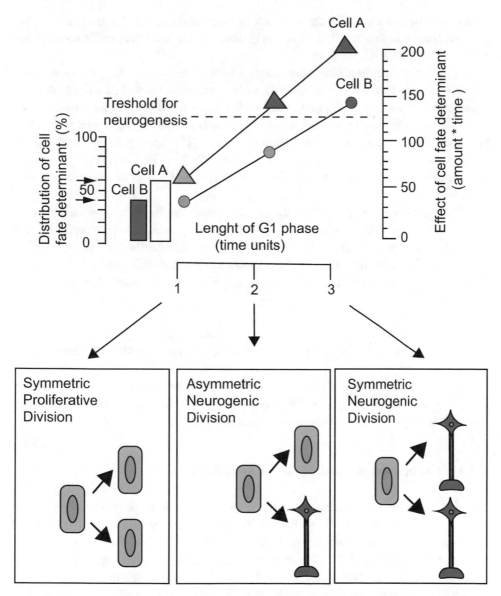

Fig. 9.3 The cell-cycle length hypothesis. A neurogenic cell fate determinant that, following cell division, is distributed unequally to daughter cells A and B (60% and 40%, respectively) can induce one or both of the daughter cells to become a neuron depending on whether G1 phase is sufficiently long for the cell fate determinant to achieve its neurogenic effect. So, neither cell A nor cell B will become a neuron after one unit of time. Cell A, but not cell B, will become a neuron after two units of time. Both cell A and cell B will become a neuron after three units of time. Redrawn from [29, 31]

Many intrinsic and extrinsic factors are responsible for regulating the generation of this patterned neuronal structure. In humans, as in most mammals, corticogenesis is completed before or around the time of birth.

An intriguing question is whether defined neural progenitors always only generate neurons for a specific layer or whether the neural progenitors would give rise to neurons populating different layers. Elegant clonal analysis suggests that the latter scenario is the case and hence that also mammalian radial glial cells commit to a temporal program to produce different neuronal subpopulation over time [33]. Transplantation experiments show that neural progenitors become restricted with time in a way that progenitors generating upper-layer neurons cannot revert back and give rise to the deep-layer (older) neurons [34].

A temporal order of progeny cells was also observed in the vertebrate retina. In chicken, Xenopus or fish ganglion cells, horizontal cells, cones and amacrine cells seem to differentiate first whereas bipolar rods and Müller glia differentiate last. However, more recent clonal analysis and mathematical modeling suggest that retinogenesis in vertebrates follows a stochastic model, in which division mode and cell fates cannot be predicted by birth order only [35].

9.2   Neural Stem Cell and Progenitor Cell Types in the Neocortex

The layered structure and the columnar architecture of the neocortex are evolutionarily conserved in mammalian species. During mammalian evolution, however, a substantial morphological transformation happened in primates [36, 37].

9.2.1   Lissencephalic and Gyrencephalic Brains

The surface of the mouse neocortex is smooth, which is a characteristic described as lissencephalic. In contrast, the surface of the human neocortex is highly convoluted and displays a pattern of sulci (fissures) and gyri (ridges). Hence, the human neocortex is described as gyrencephalic. The pattern of gyrification correlates with the enlarged neocortex in primates through radial expansion. The folded structure greatly increases the surface and therefore primate brains can contain a higher number of neurons than any other mammalian species. During evolution, it appears that the neocortex is enlarged disproportionately as compared with most other compartments of the brain [36, 37].

9.2.2   Populations of Stem and Progenitor Cell Types in the Neocortex

Advances in cell labeling techniques and live-cell microscopy in brain sections greatly contributed to the identification and characterization of neural stem and progenitor cell

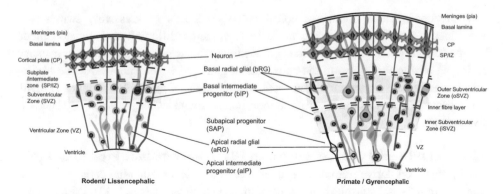

Fig. 9.4 Coronal section of developing neocortex from a representative lissencephalic species, such as mouse or rat (left), and a representative gyrencephalic species, such as ferret or human (right), showing the neural progenitor cell types observed in each of the germinal zones. Redrawn from [39]

types in lissencephalic and gyrencephalic brains [38]. The differences in neocortex structure and size are also reflected in the composition of different population of neural progenitor cells (Fig. 9.4). In the developing mammalian neocortex three main classes of neural progenitor cell types can be found: apical progenitors, basal progenitors, and subapical progenitors. We briefly describe each of these types and how they contribute to the variation in brain development [39].

9.2.2.1 Neuroepithelial Cells

Neuroepithelial cells derive from the neuroectoderm and initially form a columnar monolayer building the neural plate. In the early neural tube, the epithelium becomes pseudostratified. Neuroepithelial cells rapidly expand the progenitor pool through initial symmetric divisions. It leads to the lateral growth of the neocortex. Neuroepithelial cell growth serves also to expand in the radial dimension and to thicken the neuroepithelium. Neuroepithelial cells are highly polarized cells that span the entire developing neural tube facing the ventricular lumen at the apical side and reaching to the basal lamina. As mentioned earlier one typical characteristic of neuroepithelial cells is the interkinetic nuclear migration whereby the nuclei are found at different positions along the apical-basal axis depending on the cell cycle phase [36, 40].

9.2.2.2 Apical Radial Glial Cells (aRG) and Apical Intermediate Progenitors (aIP)

Neuroepithelial cells will eventually transform into apical radial glial cells, which constitutes the major neural stem cell type during neurogenesis [41]. Radial glial cells possess apical-basal polarity, undergo interkinetic nuclear migration, and express the transcription factor Pax6.

While neurogenesis continues, apical radial glial cells progressively switch from a symmetric proliferative division mode to asymmetric differentiative division mode. Neurons can be generated either directly or via intermediate progenitor cells. For instance, during an asymmetric division the apical radial glial cell is self-renewed while also an apical intermediate progenitor is generated. While the radial glial cell continues to divide the intermediate progenitor loses its mitotic potential and undergoes limited rounds of divisions [31].

9.2.2.3 Basal Radial Glial Cells (bRG) and Basal Intermediate Progenitors (bIP)

Basal progenitors derive from neuroepithelial cells or apical radial glial cells and delaminate toward the basal subventricular zone of the developing cortex [42, 43]. Primary basal radial glial cells (bRG) or also called outer radial glial cells (oRG) are generated apically but on their way towards the subventricular zone lose the apical connection. In primates bRG constitute an entire new zone the outer subventricular zone (oSVZ). They can self-renew through asymmetric or symmetric divisions and thereby generating secondary bRG and intermediate progenitors. Like apical-basal progenitors, they express Pax6.

In contrast, basal intermediate progenitors downregulate radial glial cell markers and start to express the transcription factor Tbr2. They can undergo a self-consuming symmetric division in the first mitotic round or after an additional proliferative symmetric division.

9.2.3   Cellular Mechanisms of Neocortex Expansion

The human brain is characterized by a highly convoluted cortex with an expanded surface area. What are the differences between a lycencephalic mammalian cortex such as for example in *Mus musculus*, a gyrencephalic non-human primate cortex such as for example *Macaca mulatta* and a highly expanded human cortex as in *Homo sapiens*? What makes the human brain so special and equips humans with such cognitive abilities. More recent observations in the primate brain gave insight into the cellular mechanisms that were potentially used during millions of years in evolving the human neocortex [37, 44].

As described above, a first stage before the onset of cortical neurogenesis neuroepithelial cells undergo sequences of symmetric divisions. It leads to the exponential increase of the number of developmental stem cells before a switch to asymmetric neurogenic divisions occurs. During neurogenesis a larger initial pool of founder cells can be converted into intermediate or terminally differentiated neurons and glial cells. A difference of just seven extra rounds of proliferative symmetric division during early embryonic stages can account for a 1000-fold difference in total cortical surface area between mice and humans.

Asymmetric self-renewing division of apical progenitor cells results in basal progenitor cell such as basal radial glial cells and intermediate basal progenitor cells (Fig. 9.5). It is assumed that these basal progenitors that populate the outer subventricular zone are the primary basis of neocortex expansion in the radial dimension. The increase in

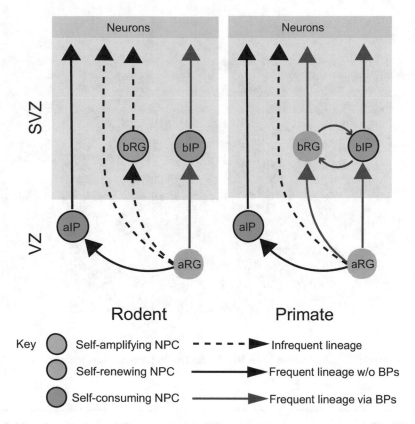

Fig. 9.5 Neural progenitor cell lineage relationships in the developing neocortex. Lineage relationships of neural progenitor cells in the developing neocortex of a representative lissencephalic rodent (left) and a gyrencephalic primate (right). Neurons (gray box) represent the end point of each neural progenitor cell lineage. Preferential modes of neural progenitor cell divisions are color-coded. Neural progenitor cells that are capable of both self-renewing and proliferative divisions are colored both pink and blue. Arrows indicate lineage progression as detailed in the key. Redrawn from [39]

diversity and population size of basal progenitor also leads to an altered shape of radial units from cylindrical to conical, with the narrower end at the ventricle. Indeed, in developing mammalian neocortex with a high gyrification index such as monkey and humans, proliferating basal progenitor cells are particularly abundant [39, 42].

Question
Discuss the morphological and cellular mechanisms that lead to differences in brain cell numbers between lower mammals and primates.

Take-Home Message

- Mammalian neurogenesis follows a more stochastic pattern as compared to stereotypic neural stem cell lineages observed in *Drosophila*.
- Delta-Notch signaling maintains and oscillating proneural gene expression maintains the undifferentiated progenitor state, while sustained proneural gene expression promotes neurogenesis and differentiation.
- Neurogenesis of the layered mammalian neocortex follows a spatio-temporal "inside first and outside last" pattern.
- Neuroepithelial cells and radial glial cells are the major neural stem cell types during lateral and radial expansion phase of the mammalian cerebral cortex.
- Convoluted or gyrencephalic brain architecture allows higher neuronal number and complexity among primate brains.
- Higher evolved gyrencephalic brains reveal a higher number of basal or outer radial glial cells that constitute an outer subventricular zone.

Acknowledgment We are grateful to Dr Rita Sousa-Nunes for comments and suggestions for this chapter.

References

1. Chenn A, McConnell SK. Cleavage orientation and the asymmetric inheritance of Notch1 immunoreactivity in mammalian neurogenesis. Cell. 1995;82(4):631–41.
2. McConnell SK. Constructing the cerebral cortex: neurogenesis and fate determination. Neuron. 1995;15(4):761–8.
3. Miyata T, Kawaguchi D, Kawaguchi A, Gotoh Y. Mechanisms that regulate the number of neurons during mouse neocortical development. Curr Opin Neurobiol. 2010;20(1):22–8.
4. McConnell SK. The determination of neuronal identity in the mammalian cerebral cortex. In: Shankland M, Macagno ER, editors. Determinants of neuronal identity. London: Academic; 1992.
5. Noctor SC, Flint AC, Weissman TA, Wong WS, Clinton BK, Kriegstein AR. Dividing precursor cells of the embryonic cortical ventricular zone have morphological and molecular characteristics of radial glia. J Neurosci. 2002;22(8):3161–73.
6. Malatesta P, Hack MA, Hartfuss E, Kettenmann H, Klinkert W, Kirchhoff F, et al. Neuronal or glial progeny: regional differences in radial glia fate. Neuron. 2003;37(5):751–64.
7. Hartfuss E, Galli R, Heins N, Götz M. Characterization of CNS precursor subtypes and radial glia. Dev Biol. 2001;229(1):15–30.
8. Lancaster MA, Knoblich JA. Spindle orientation in mammalian cerebral cortical development. Curr Opin Neurobiol. 2012;22(5):737–46.
9. Paridaen JT, Huttner WB. Neurogenesis during development of the vertebrate central nervous system. EMBO Rep. 2014;15(4):351–64.

10. Konno D, Shioi G, Shitamukai A, Mori A, Kiyonari H, Miyata T, et al. Neuroepithelial progenitors undergo LGN-dependent planar divisions to maintain self-renewability during mammalian neurogenesis. Nat Cell Biol. 2008;10(1):93–101.

11. Postiglione MP, Juschke C, Xie Y, Haas GA, Charalambous C, Knoblich JA. Mouse inscuteable induces apical-basal spindle orientation to facilitate intermediate progenitor generation in the developing neocortex. Neuron. 2011;72(2):269–84.

12. Marthiens V, Basto R. Centrosomes: the good and the bad for brain development. Biol Cell. 2020;112(6):153–72.

13. Thornton GK, Woods CG. Primary microcephaly: do all roads lead to Rome? Trends Genet. 2009;25(11):501–10.

14. Lee JE. Basic helix-loop-helix genes in neural development. Curr Opin Neurobiol. 1997;7(1):13–20.

15. Bertrand N, Castro DS, Guillemot F. Proneural genes and the specification of neural cell types. Nat Rev Neurosci. 2002;3(7):517–30.

16. Casarosa S, Fode C, Guillemot F. *Mash1* regulates neurogenesis in the ventral telencephalon. Development. 1999;126(3):525–34.

17. Ma Q, Fode C, Guillemot F, Anderson DJ. *Neurogenin1* and *Neurogenin2* control two distinct waves of neurogenesis in developing dorsal root ganglia. Genes Dev. 1999;13(13):1717–28.

18. Fode C, Ma Q, Casarosa S, Ang SL, Anderson DJ, Guillemot F. A role for neural determination genes in specifying the dorsoventral identity of telencephalic neurons. Genes Dev. 2000;14(1):67–80.

19. Ma Q, Kintner C, Anderson DJ. Identification of *neurogenin*, a vertebrate neuronal determination gene. Cell. 1996;87(1):43–52.

20. Beatus P, Lendahl U. *Notch* and neurogenesis. J Neurosci Res. 1998;54(2):125–36.

21. Chitnis A, Henrique D, Lewis J, Ish-Horowicz D, Kintner C. Primary neurogenesis in *Xenopus* embryos regulated by a homologue of the *Drosophila* neurogenic gene *Delta*. Nature. 1995;375(6534):761–6.

22. Coffman CR, Skoglund P, Harris WA, Kintner CR. Expression of an extracellular deletion of Xotch diverts cell fate in *Xenopus* embryos. Cell. 1993;73(4):659–71.

23. Wettstein DA, Turner DL, Kintner C. The *Xenopus* homolog of *Drosophila Suppressor of Hairless* mediates Notch signaling during primary neurogenesis. Development. 1997;124(3):693–702.

24. de la Pompa JL, Wakeham A, Correia KM, Samper E, Brown S, Aguilera RJ, et al. Conservation of the Notch signalling pathway in mammalian neurogenesis. Development. 1997;124(6):1139–48.

25. Pierfelice TJ, Schreck KC, Eberhart CG, Gaiano N. Notch, neural stem cells, and brain tumors. Cold Spring Harb Symp Quant Biol. 2008;73:367–75.

26. Kageyama R, Ohtsuka T, Shimojo H, Imayoshi I. Dynamic Notch signaling in neural progenitor cells and a revised view of lateral inhibition. Nat Neurosci. 2008;11(11):1247–51.

27. Imayoshi I, Isomura A, Harima Y, Kawaguchi K, Kori H, Miyachi H, et al. Oscillatory control of factors determining multipotency and fate in mouse neural progenitors. Science. 2013;342(6163):1203–8.

28. Shimojo H, Ohtsuka T, Kageyama R. Oscillations in notch signaling regulate maintenance of neural progenitors. Neuron. 2008;58(1):52–64.

29. Calegari F, Huttner WB. An inhibition of cyclin-dependent kinases that lengthens, but does not arrest, neuroepithelial cell cycle induces premature neurogenesis. J Cell Sci. 2003;116(Pt 24):4947–55.

30. Lange C, Huttner WB, Calegari F. Cdk4/cyclinD1 overexpression in neural stem cells shortens G1, delays neurogenesis, and promotes the generation and expansion of basal progenitors. Cell Stem Cell. 2009;5(3):320–31.
31. Gotz M, Huttner WB. The cell biology of neurogenesis. Nat Rev Mol Cell Biol. 2005;6(10):777–88.
32. Bonnefont J, Vanderhaeghen P. Neuronal fate acquisition and specification: time for a change. Curr Opin Neurobiol. 2021;66:195–204.
33. Gao P, Postiglione MP, Krieger TG, Hernandez L, Wang C, Han Z, et al. Deterministic progenitor behavior and unitary production of neurons in the neocortex. Cell. 2014;159(4):775–88.
34. Desai AR, McConnell SK. Progressive restriction in fate potential by neural progenitors during cerebral cortical development. Development. 2000;127(13):2863–72.
35. He J, Zhang G, Almeida AD, Cayouette M, Simons BD, Harris WA. How variable clones build an invariant retina. Neuron. 2012;75(5):786–98.
36. Rakic P. Evolution of the neocortex: a perspective from developmental biology. Nat Rev Neurosci. 2009;10(10):724–35.
37. Lui JH, Hansen DV, Kriegstein AR. Development and evolution of the human neocortex. Cell. 2011;146(1):18–36.
38. Breunig JJ, Haydar TF, Rakic P. Neural stem cells: historical perspective and future prospects. Neuron. 2011;70(4):614–25.
39. Florio M, Huttner WB. Neural progenitors, neurogenesis and the evolution of the neocortex. Development. 2014;141(11):2182–94.
40. Rakic P. A small step for the cell, a giant leap for mankind: a hypothesis of neocortical expansion during evolution. Trends Neurosci. 1995;18(9):383–8.
41. Malatesta P, Hartfuss E, Götz M. Isolation of radial glial cells by fluorescent-activated cell sorting reveals a neuronal lineage. Development. 2000;127(24):5253–63.
42. Hansen DV, Lui JH, Parker PR, Kriegstein AR. Neurogenic radial glia in the outer subventricular zone of human neocortex. Nature. 2010;464(7288):554–61.
43. Fietz SA, Kelava I, Vogt J, Wilsch-Brauninger M, Stenzel D, Fish JL, et al. OSVZ progenitors of human and ferret neocortex are epithelial-like and expand by integrin signaling. Nat Neurosci. 2010;13(6):690–9.
44. Taverna E, Gotz M, Huttner WB. The cell biology of neurogenesis: toward an understanding of the development and evolution of the neocortex. Annu Rev Cell Dev Biol. 2014;30:465–502.

Models of Neurodegenerative Diseases

10

Niran Maharjan, Smita Saxena

What You Will Learn in This Chapter
Neurodegenerative diseases comprise a wide range of age-related conditions, characterized by the loss of neurons leading to a progressive decline in brain function. Diverse cellular and animal models have enhanced our understanding of the molecular pathogenesis of different neurodegenerative diseases. However, failure to translate potential therapeutic findings from bench to bedside is primarily due to limitations of animal models and needs urgent attention. In this chapter, we discuss different neurodegenerative diseases and compare the different models commonly used in their study.

10.1 Introduction

Neurodegenerative disease is an umbrella term given to conditions wherein a gradual loss of neurons in the central nervous system results in cognitive and behavioral deficits. Monogenic forms of neurodegenerative diseases are typically characterized by the expression of specific mutant proteins and are heterogeneous in their clinical manifestations due to specific anatomic vulnerability. However, most of them often have overlapping neuropathological features and pathomechanisms. A common feature of neurodegenera-

N. Maharjan · S. Saxena (✉)
Department of Neurology, Inselspital University Hospital, Bern, Switzerland

Department for BioMedical Research, University of Bern, Bern, Switzerland
e-mail: niran.maharjan@dbmr.unibe.ch; smita.saxena@dbmr.unibe.ch

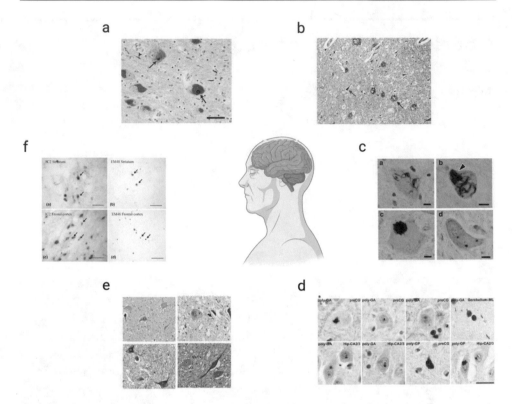

Fig. 10.1 Presence of protein aggregation and inclusions is a pathological hallmark of different neurodegenerative diseases. (**a**) Lewy body (arrow) and Lewy neurites (arrow head) in PD patients [2]. (**b**) Plaques (black arrow) and neurofibrillary tangles (red arrow) in AD patients [3]. (**c**) Different patterns of TDP-43 aggregates in ALS patients [4]. (**d**) Different dipeptide repeats (poly GA and poly GP) in different parts of the brain from ALS patients with C9ORF72 repeat expansion [5]. (**e**) misfolded SOD1 accumulation in familial ALS patients [6]. (**f**) Aggregated huntingtin observed in HD patients [7]

tive diseases is the deposition of insoluble protein aggregates formed by disease-specific proteins. These proteins include β-amyloid (Aβ) and tau in Alzheimer's disease (AD), α-synuclein in Parkinson's disease (PD), dementia with Lewy bodies (DLB) and multiple system atrophy (MSA), TDP-43 in amyotrophic lateral sclerosis (ALS) and frontotemporal lobar degeneration (FTLD) and polyglutamine (PolyQ) rich inclusions in Huntington's disease (HD) and Spinocerebellar ataxias (SCA) [1] (Fig 10.1).

Although the familial forms of neurodegenerative diseases are rarer than the sporadic forms, the pathophysiological mechanisms are thought to be conserved. It is well acknowledged that the fundamental mechanism behind neurodegenerative diseases is polyfactorial and depends on the complex interplay of multiple genetic and nongenetic variables [8]. Although most of the neurodegenerative disease cases have no clear genetic links, the field has been steered by the discovery of different mutant genes or risk factor genes that drive these disorders. Identification of genetic mutations makes it possible to use genetic manipulation to model these diseases in different animals. A variety of genetically modified organisms, ranging from yeast to rodents, have been generated to study the disease pathogenesis and therapeutic targets [9]. Although different models have provided important insights into understanding the pathogenic mechanism leading to brain cell dysfunction and degeneration, not all models fully mimic the symptoms and pathogenesis of neurodegenerative diseases. Hence, a multiplicity of models is required to better understand neurodegenerative diseases. The current chapter will provide an overview of different models used in AD, PD, and ALS.

10.1.1 Alzheimer's Disease (AD)

Alzheimer's disease is an adult-onset neurodegenerative disease that mainly affects people over 60 years and is characterized by early progressive memory loss, executive function impairments, and personality disorder. AD is the most common cause of dementia, responsible for 60–80% of cases [10]. In 2020, an estimated about 6 million Americans aged 65 or over had Alzheimer's disease and this number is estimated to grow to 14 million by 2060 [11]. The rate of progression for AD varies widely from person to person. On average, patients with AD live between 3 and 10 years after diagnosis, while some survive more than 20 years. Clinically, AD progression is divided into different stages, mild AD, moderate AD, and severe or advanced AD [12]. AD begins with a gradual decline in memory and slowly progresses in severity. With disease progression, impairment in other areas of cognition occurs, affecting cognitive, social, and intellectual abilities. The majority of the AD cases are sporadic, while only a few cases (about 10%) are familial, linked to a genetic mutation in Presenilin 1 (PSEN1), Presenilin 2 (PSEN2), and Amyloid precursor protein (APP) [12] (Fig. 10.2).

The extracellular amyloid plaques and neurofibrillary tangles are the neuropathological hallmarks of AD. The major components of amyloid plaques consist of misfolded Aβ

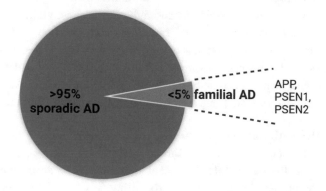

Fig. 10.2 Pie chart showing the prevalence of sporadic and familial AD in percentage. Indicated are genes associated with familial AD

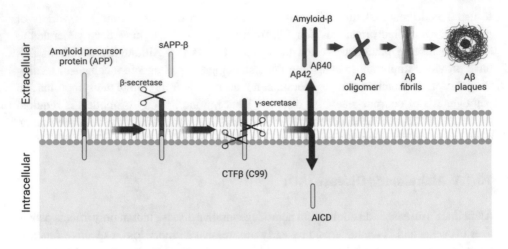

Fig. 10.3 Schematic representation of amyloid precursor protein processing and amyloid plaque formation

sheets. Aβ deposits are derived from the proteolytic cleavage of the transmembrane amyloid precursor protein (APP). At first, β-secretase cleaves APP resulting in APP-β and the membrane-bound C99 fragment. The cleaved APP fragment (C99) is further cleaved by γ-secretase to produce Aβ40 or Aβ42 fragments depending on the point of cleavage [13]. Aβ40 and Aβ42 fragments aggregate into oligomers and fibrils, resulting in amyloid plaque formation (Fig. 10.3). Although Aβ aggregation is essential for initiating the AD cascade, it is not sufficient to cause AD pathogenesis as there is minimal correlation between amyloid deposition and degree of cognitive decline [14]. Instead, the amyloid cascade hypothesis states that the deposition of Aβ aggregate is the instigating step of AD pathology, leading to subsequent neurofibrillary tangles, neuronal loss, and cognitive decline [15]. Assessment of progressive accumulation of amyloid plaques and tau neurofibrillary tangles (NFT) using positron emission tomography (PET) imaging on AD patients validates the amyloid cascade hypothesis, as amyloid accumulation was found to precede tau accumulation and the rate of tau accumulation predicts the onset of cognitive impairment [16]. Tau protein is encoded by the Microtubule Associated Protein Tau (MAPT) gene and is primarily expressed in neurons in CNS. Tau plays an important role in microtubule assembly and stabilization of neuronal axons. In AD, tau protein is abnormally hyperphosphorylated, and aggregates to form neurofibrillary tangles.

10.1.1.1 Models of Alzheimer's Disease

Different animal models have been used to understand the disease pathophysiology of AD. To date, more than 200 different rodent models have been developed and used to recapitulate the human AD pathology in rodent models [17]. Most genetically-based rodent models rely on producing either amyloid or tau pathology by expressing AD-linked human mutation (Table 10.1). The first transgenic AD model was developed by

Table 10.1 Common mouse models used in AD research

Mouse model of AD	Gene/mutation	Promoter	Neuropathology	Cognitive deficits	Neuron loss	Ref.
NSE-APP751	APP751 isoform	Neuronal specific enolase (NSE) promotor	Aβ deposits visible by 2 months but does not form mature plaque. Although aberrant tau immunoreactivity was observed by 2 months no classical NFT was observed.	Yes (12 months)	No	[18]
PDAPP	APP V717F (Indiana)	PDGF	Aβ deposits visible by 6–9 months and plaque become extensive with age. No NFT visible but at 14 months phosphorylated tau visible.	Yes (3 months)	No	[20, 27]
Tg2576	APP K670M/N671L (Swedish)	Hamster Prion Protein	Aβ plagues by 11–13 months No NFT	Yes (6 months)	No	[21]
APP23	APP K670M/N671L (Swedish)	Mouse Thy1	Aβ deposits visible at 6 months and increase in size with age. Dystrophic neurites containing hyperphosphorylated tau but absence of NFT	Yes (3 months)	Yes	[22]
3xTg	APP K670M/N671L (Swedish), MAPT P301L, PSEN1 M146V	Mouse Thy1.2	Extracellular Aβ deposits by 6 months Tau pathology by 12 months	Yes (4 months)	No	[28]
5xFAD (B6SJL)	APP K670M/N671L (Swedish), APP I716V (Florida), APP V717I (London), PSEN1 M146L, PSEN1 L285V	Mouse Thy1	Amyloid pathology starts at 2 months Absence of NFT	Yes (4–5 months)	Yes	[25]

overexpressing human wild-type APP751 isoform under the rat neuronal specific enolase (NSE) promotor. Although this model had increased Aβ deposits, it exhibited neither classic mature plaques nor neurofibrillary tangles [18, 19]. Later, to mimic human AD pathogenesis, multiple transgenic mice carrying human APP with familial AD-associated mutations were generated. PDAPP mice expressed human APP with the Indiana mutation (APP V717F) under the PDGF-β promotor [20]; while Tg2576 [21] and APP23 [22] mice expressed human APP with the double Swedish mutation (APP K670N/M671L) either under a prion protein promoter (PrP) or a neuron-specific Thy1 promotor. The Swedish mutations increase APP processing, resulting in increased intracellular Aβ level compared to the wild-type control; while the Indiana mutation results in an increased ratio of Aβ40/Aβ42 [23]. PSEN1/2 mutation is also known to alter APP processing resulting in an increased ratio of Aβ40/Aβ42 [24]. Mouse model, 5XFAD, with both APP and PSEN1/2 displayed severe amyloid phenotype due to an increase in Aβ42 production. These mice reliably recapitulate the majority of the AD pathology seen in human AD patients, including plaque formation, dystrophic neurites, microgliosis, neuronal loss, and behavioral and motor deficits along with memory impairments [25, 26]. Although these mice lack neurofibrillary tangle pathology, they are widely used as a model for Aβ-induced amyloid plaque formation and neurodegeneration.

Neurofibrillary tangles are seen in transgenic mice expressing human tau associated with FTLD mutations (P301L) [29, 30]. These mouse models have limitations in AD research, as they are more reliant on FTD mutations, which are not associated with pure forms of AD. However, co-expression of mutated versions of APP, MAPT, and PSEN1/2 resulted in the development of both amyloid plaque and neurofibrillary tangles in the same model. 3xTG-AD is a widely used mouse model, which contains three different mutations (APP Swedish, PSEN1 M146V, and tau P301L). This model is considered to be the complete AD mouse model as it exhibits both Aβ and tau pathology that is characteristic of human AD. Development of both plaques and tangles are age-dependent, whereby extracellular Aβ deposits are apparent by 6 months in the frontal cortex and become extensive with age, whereas tau pathology becomes evident by 12 months. These mice also show synaptic impairments, cognitive deficits, neuronal loss and memory deficits, which are a major hallmark of AD [28, 31]. Although the 3xTg mouse model recapitulates most AD pathology, it is not a perfect model for AD as the levels of Aβ are non-physiological and rely on FTD-associated tau mutation to generate the tauopathy.

Although mice are evolutionarily closer to humans than other model systems, there are considerable differences between humans and mice in terms of protein sequences and their function. For example, in mice, the APP protein sequence differs from human APP by 17 amino acids, of which 3 (R5G, Y10F, and H13R) are located within the N terminal domains of Aβ peptide and impair its aggregation propensity [32]. Similarly, tau isoform profiles are also different in humans and mice: humans have both 3R and 4R isoforms while adult mice express only 4R isoforms [33]. Hence, AD mouse models essentially require the overexpression of human proteins with familial AD mutations at much higher levels than the physiological state required to develop AD phenotype.

10.1.2 Parkinson's Disease (PD)

Parkinson's disease is the second most common neurodegenerative disease after AD and is characterized by the progressive loss of dopamine (DA) neurons in the substantia nigra pars compacta (SNpc), resulting in motor symptoms including bradykinesia, postural instability, and resting tremor [34]. The etiology of PD is complex involving both genetic and environmental factors. Although most PD cases are sporadic, around 10–15% of the cases are linked to different genes and are known as familial PD. Mutations in different genes including α-synuclein (SNCA), leucine repeat kinase 2 (LRRK2), parkin (PRKN), PTEN-induced putative kinase 1 (PINK1), and DJ-1 are linked with familial PD (Fig. 10.4). Although the exact cause of sporadic PDs is still unknown, to date more than 40 PD risk loci, including GBA and UCHL1, have been found to modify the risk of developing PD [34].

α-synuclein is the major component of Lewy bodies and is a pathological hallmark of familial and sporadic PD. Other neurodegenerative diseases like dementia with Lewy bodies and multiple system atrophy (MSA) also contain α-synuclein positive inclusions in neurons and glial cells and are collectively termed as synucleinopathies [35]. Under normal conditions, 140 amino acid α-synuclein exists in a dynamic equilibrium between unfolded monomer and α-helical folded tetramer in neurons and plays an important role in neuronal plasticity. However, different triggers, including point mutations or overproduction of the protein, result in the accumulation and aggregation of α-synuclein. During aggregation, soluble α-synuclein monomers undergo conformational change whereby they adopt a β-sheet-rich structure and form α-synuclein fibrils. These fibrils are generated by a nucleation-dependent mechanism, initially forming oligomers, and then progressively combining to form small protofibrils, eventually resulting in mature fibrils, which accumulate in Lewy bodies [36]. According to the Braak hypothesis, the progression of α-synuclein pathology follows a particular pattern following a spatiotemporal progression

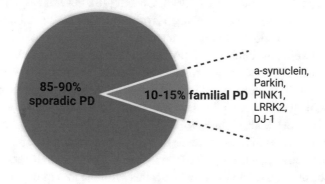

Fig. 10.4 Prevalence of sporadic and familial Parkinson's disease (PD) depicted via a pie chart. Representative genes linked to familial PD

[37]. Observation of α-synuclein pathology in grafted fetal mesencephalic progenitor neurons, several years after transplantation in PD patients, provided evidence for the transmission of Lewy body pathology from the host to the graft, reminiscent of prion-like propagation [38, 39].

10.1.2.1 Models of Parkinson's Disease

Given the important role of propagation and aggregation of α-synuclein in PD pathogenesis, one of the approaches to create a cellular or mouse model has been to use either preformed fibrils (PFFs) generated in vitro from recombinant α-synuclein or the brain extract from PD patients. Internalized PFFs work as a template to recruit endogenous α-synuclein resulting in its aggregation with different cellular components, and leading to the formation of Lewy body-like inclusions in neuronal cultures [40, 41]. The PFF model has since been extended to rodent models wherein intracerebral injection of α-synuclein PFF resulted in α-synuclein aggregation resembling Lewy body pathology. Importantly, the aggregated α-synuclein colocalized with common markers of human Lewy bodies including p62, ubiquitin, and phosphorylated α-synuclein, thus replicating human pathological features. These models also show the spreading of α-synuclein pathology to anatomically interconnected brain regions from the injection site. PFF-injected mice also show progressive degeneration of dopaminergic neurons, neuroinflammation, and motor deficits without the need for overexpressing α-synuclein beyond physiological levels [42–45]. However, the concentration and strain of the injected PFF should be taken into consideration as α-synuclein PFF is heterogenous in nature, and rodents injected with different PFF strains manifest distinctive pathology and neurotoxic phenotypes [46, 47].

Given the heterogeneity of PD, a wide range of genetic and neurotoxin approaches has been used to model PD in rodents. Most of the transgenic rodent models are generated by targeting one of the five genes (SNCA, LRRK2, PRKN, PINK1, and DJ-1) associated with familial PD (Table 10.2). Mutations in α-synuclein are linked to autosomal dominant PD and are a major Lewy body pathology component [48]. Several α-synuclein transgenic mice have been developed overexpressing α-synuclein with PD-linked mutations (A53T, A30P, E46K) under various promoters including PrP and Thy-1. Although different mouse models develop the clinical and biochemical features of PD, including α-synuclein aggregation, reactive gliosis, and neuronal degeneration, most of the models fail to show loss of dopaminergic neurons. While fibrillar α-synuclein aggregates are seen in most mouse models, they do not resemble Lewy body [49, 50]. Surprisingly, mice expressing α-synuclein harboring both mutations (A53T and A30P) show a progressive loss of dopaminergic neurons even in the absence of α-synuclein inclusions [51]. However, mice expressing truncated α-synuclein (1–120) on mouse α-synuclein null background resulted in the formation of α-synuclein inclusions together with the loss of dopaminergic neurons [52].

Mutations in the LRRK2 gene are linked to both familial and sporadic PD and account for 1% of all PD cases. Several LRRK2-related PD rodent models have been generated

Table 10.2 Common mouse model used in PD research

Mouse model of PD	Gene/mutation	Promoter	Dopamine deficiency	Neuropathology	Motor impairment	Neuron loss	Ref.
α-synuclein A53T (Tg)	SNCA A53T	Mouse prion protein	At 5 months	Accumulation of fibrillar α-synuclein, ubiquitin and neurofilament	Severe motor impairment starting around 9 months	No	[49]
Thy1-αSyn "Line 61" mice	SNCA	Mouse Thy1	At 14 months	Proteinase K resistant aggregates seen as early as 1 month and increase with age. Increase in pS129 α-synuclein level	Impairments in balance, coordination arise from 1 to 3 months	Loss of neurons in neocortex and hippocampus as early as 3–4 months. Even at 22 months no visible loss of dopaminergic neurons	[53]
α-synuclein A30P/A53T mouse (Tg)	SNCA A30P, SNCA A53T	Tyrosine hydroxylase (TH) promoter	Lower striatal dopamine level as early as 2 months	Presence of diffused α-synuclein in cytoplasm and nucleus. Inclusion not observed at any age	Reduced motor coordination starting at 13 months	Loss of dopaminergic neurons starting at 9 months	[51]
α-synuclein A53T mouse (Tg) on SNCA KO	SNCA A53T KO of mouse α-synuclein	Expression of Human α-synuclein A53T with via P1 artificial chromosome (PAC) KO of mouse SNCA	No difference	No visible α-synuclein aggregates	Motor impairment visible from 6 months.	No neuronal loss	[50, 54]

(continued)

Table 10.2 (continued)

Mouse model of PD	Gene/mutation	Promoter	Dopamine deficiency	Neuropathology	Motor impairment	Neuron loss	Ref.
Parkin Q311X Mouse (BAC Tg)	Parkin Q311X	Bacterial artificial chromosome (BAC) mediated expression of Parkin Q311X in dopaminergic neurons through promotor of Slc6a3	Lower level of dopamine by 19 months	Age-dependent accumulation of proteinase K resistant α-synuclein. No inclusion formation at any age	Deficits in coordination by 16 months	40% of dopaminergic loss by 16 months	[55]
PINK1 KO mouse		Knock out of PINK1 gene	No significant decrease in dopamine level	No data on synuclein pathology	Reduced locomotor activity at 3–6 months	No loss of dopaminergic neurons	[56]
PINK1 G309D (PINK1-/-) mouse (KI)	Pink1 G309D	G309D mutation introduced into exon 4 at the orthologous mouse locus via homologous recombination	Decrease in dopamine concentration at 9 months	Although there is altered expression of α-synuclein, no Lewy body-like inclusion observed	Motor impairment observed at 16 months	No neuronal loss	[57]
LRRK2 R1441C mosuse (Tg-conditional)	LRRK2 R1441C	Human LRRK2 R1441C targeted to endogenous ROSA26 locus by homologous recombination	No significant difference in Dopamine level	No α-synuclein abnormalities, No inclusion formation	No motor impairment	No neuronal loss	[58]
LRRK2 G2019S mouse (Tg)	LRRK2 G2019S	CMVe-PDGFβ	No significant difference in dopamine levels	No visible α-synuclein pathology	Decline in Rotarod performance starting from 14 months	About 18% of dopaminergic neurons are lost by 19 months	[59]
LRRK2 G2019S mouse (BAC Tg)	LRRK2 G2019S	BAC containing mouse LRRK2 G2019S with Lrrk2 promoter region	Decrease in dopamine at 12 months	No visible α-synuclein inclusion	No visible motor impairment	No neuronal loss	[60]

by overexpressing either wild-type or pathogenic variant of LRRK2. Transgenic mice overexpressing LRRK2 mutants (G2019S/ R1441C) show PD-like phenotypes, including the loss of dopaminergic neurons, levodopa-responsive (dopamine precursor) locomotor deficits, and pathological accumulation of α-synuclein [59, 61]. Likewise, mice with Q311X mutation in the gene parkin result in C-terminally truncated Parkin expression and show loss of dopaminergic neurons with progressive changes in motor behavior with age. Q311X mice do not develop Lewy body pathology, but show the accumulation of α-synuclein in substantia nigra [55]. Notably, none of the genetic PD models can entirely recapitulate the human PD pathology, but their use has given us valuable information on different aspects of PD.

Since PD in most patients does not have a clear genetic cause, various mouse model setups based on neurotoxins are efficient in reproducing dopamine deficit. The most widely used neurotoxins are 1-methyl-4-phenyl-1,2,3,6-tetrahydropyridine (MPTP) and 6-hydroxy-dopamine (6-OHDA). Once MPTP reaches the brain by crossing the blood-brain barrier (BBB), it is metabolized into MPP+ and induces neuronal toxicity via mitochondrial dysfunction [62]. Although 6-OHDA cannot cross BBB, when it is injected inside the brain it is broken down to form hydrogen peroxide and free radicals and causes toxicity through oxidative stress [63]. Neurotoxic models mimic many of the PD features, including reduced levels of striatal dopamine and tyrosine hydroxylase, and dopaminergic neuronal loss. Nevertheless, these models lack Lewy body pathology and therefore cannot be used to study non-dopaminergic alterations related to PD [64]. As an alternative approach, various studies have used a combinatorial strategy, whereby α-synuclein transgenic mice were administered the neurotoxin to induce dopaminergic neuronal loss along with the formation of Lewy body-like pathology [65].

10.1.3 Amyotrophic Lateral Sclerosis (ALS)

ALS is a fast progressing, fatal neurodegenerative disorder that primarily affects the neurons controlling the movement of voluntary muscles. ALS is characterized by the loss of upper and lower motor neurons leading to the development of progressive paralysis. Typical symptoms include progressive spasticity, muscle wasting, weakness, dysarthria and dysphagia [66, 67]. The clinical façade of ALS shows strong phenotypical heterogeneity with a variable mix of upper motor neuron (UMN) and lower motor neuron (LMN) degeneration. Domination of either UMN or LMN degeneration leads to significant differences in the age of onset, spreading pattern, disease severity, and progression [68]. Nevertheless, ALS is relentlessly progressing, with 50% of the patients dying within 30 months after the onset of symptoms while only about 20% of the patients survive more than 5 years. The most common reason for death in an ALS patient is respiratory failure due to the weakening of respiratory muscles, often connected with pneumonia [66].

The most striking pathological sign of ALS is the degeneration of the motor system with a massive loss of ventral horn cells along the whole spinal cord and the loss of motor neurons in the brain stem and motor cortex. In ALS, intracellular inclusions are present in degenerating neurons and glia cells. Such inclusions with misfolded and aggregated proteins strongly suggest that protein stability, misfolding, and aggregation are central to the disease progression. The identification of ubiquitinated TAR DNA-binding protein 43 (TDP-43) inclusion in the degenerating neurons consolidated the previous evidence of clinical overlap between ALS and frontotemporal dementia (FTD) [69].

Traditionally, ALS and FTD were considered two completely different neurological disorders with distinct clinical features but now are regarded as the same disease spectrum with pure motor neuron disease at one end and pure FTD at another. Following the discovery of an association between a mutation in the superoxide dismutase 1 (SOD1) gene and ALS, many studies focused on pathogenesis caused by a point mutation in SOD1 gene [70, 71]. However, the pathological findings associated with SOD1 mutations were not similar to sporadic ALS cases, suggesting the possibility of differing pathomechanism depending on ALS type [72]. In 2006, TDP-43 was discovered to be the main component of the ubiquitinated inclusion in both sporadic ALS patients and the most frequent pathological form of FTD, suggesting ALS and FTD might be part of a disease spectrum sharing a common molecular basis with TDP-43 [69, 73, 74]. Likewise, FUS-positive inclusion was also demonstrated in a subset of ALS and FTD patients with TDP-43-negative neuronal inclusion [75]. It has been established that ALS and FTD can co-occur in the same individual. Fifty to seventy-five percent of the reported ALS cases manifest some cognitive and behavioral abnormalities while approximately 15–25% of patients manifest symptoms fulfilling FTD criteria. Likewise, 40% of FTD cases have measurable motor dysfunctions [76–78]. The concept that ALS and FTD represent two ends of the same disease spectrum was further supported by genetic approaches, as several groups have detected common mutations in TDP-43, Fused in Sarcoma (FUS) and Chromosome 9 Open Reading Frame 72 (C9ORF72) responsible for both ALS and FTD [79–84].

10.1.3.1 Models of Amyotrophic Lateral Sclerosis

Although the majority of ALS cases have no family history, about 20% of ALS cases are linked to mutations in more than 20 different genes [85]. The most common genetic link is C9ORF72 which accounts for about 40% of fALS cases. Other common genes linked to ALS are SOD1, TARDBP, and FUS with frequencies of about 20%, 5%, and 4%, respectively [85] (Fig. 10.5). The identification of the genetic mutations linked to ALS was critical for ALS research. To date, different ALS models have been developed, which proved to be valuable for understanding the molecular pathogenesis linked to ALS (Table 10.3).

Mutations in SOD1 gene was the first identified cause linked to ALS in 1993, and to date more than 180 ALS-causing point mutations have been identified in the SOD1 gene [102]. SOD1 is a ubiquitously expressed protein found in the cytoplasm and mitochondria, where it acts to detoxify superoxide free radicals by converting them to oxygen and hydrogen peroxide. Initially, it was believed that mutations in the SOD1 gene cause ALS due to the inactivation of wild-type SOD1 protein, leading to a loss of function scenario.

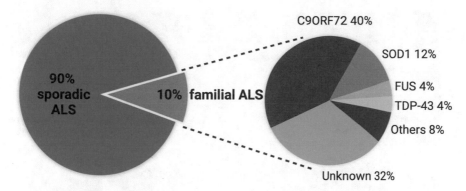

Fig. 10.5 Pie chart indicating the percentage of sporadic and familial Amyotrophic lateral sclerosis. Genes linked to familial ALS and its prevalence within ALS patients

However, SOD1 knockout mice did not produce any motor neuron phenotype similar to those observed in ALS, indicating that the mutations cause a toxic gain of function rather than the expected loss of function of the protein [103]. The universal finding is that almost all SOD1 mutants fail to fold properly, suggesting that accumulation of misfolded SOD1 is a possible toxic mechanism in ALS. The SOD-G93A transgenic mouse model was the first mouse model of ALS expressing approximately 18 copies of the human SOD1 gene-carrying G93A mutation [86]. These mice develop adult-onset degeneration of spinal motor neurons and progressive motor deficits resulting in paralysis, a hallmark of ALS. It also recapitulates many human pathological features of ALS, including mitochondrial vacuolization, cytoplasmic SOD1 aggregation, degeneration of neuromuscular junctions, gliosis, and reduced lifespan. However, the original SOD1-G93A line has diverged further into various strains carrying differing copy numbers of transgene, which significantly affects the disease onset and severity [104]. Likewise, other transgenic mouse models expressing SOD1-G85R and SOD1-G37R have been developed to study the effect of different point mutations in SOD1, and these transgenic lines also develop pathological hallmarks of ALS [87, 88]. However, a major caveat of SOD1 models is that pathologically SOD1 is distinct from other forms of ALS, and it does not recapitulate the TDP-43 pathology as observed in the majority of human ALS cases [105].

TDP-43 was first identified as the major constituent of the proteinaceous inclusion, which is characteristic of most forms of ALS and FTD. By early 2008, the first mutation in TARDBP gene was reported in ALS, providing conclusive evidence that TDP-43 itself is linked to ALS and FTD [80, 106]. Structurally related to the (heterogeneous nuclear ribonucleoprotein) hnRNP family, TDP-43 is a ubiquitously expressed 414 amino acid protein normally present in the nucleus but also shuttling back and forth between the nucleus and cytoplasm. In addition to its well-characterized role as a global regulator of gene expression, it has also been associated with multiple cellular process such as microRNA processing, apoptosis, and the stabilization of mRNA [107]. To date, more

Table 10.3 Common mouse model used in ALS research

Mouse model of ALS	Gene/mutation	Modification info	Neuropathology	Motor impairment	Muscle atrophy and neuron loss	Premature death	Ref.
SOD1-G93A	SOD1 G93A	Multiple copies of human SOD1 G93A integrated randomly into chromosome 12	SOD1 and Ubiquitin-positive, TDP-43-negative inclusion accumulate in motor neuron around 82 days	Motor impairment starts from 3 month	Reduced muscle volume. Degeneration of NMJ. Up to 50% loss of motor neuron in spinal cord at end stage	Male transgenic mice die by 5 months	[86]
SOD1-G85R	SOD1 G85R	Transgene expressing SOD1-G85R with human SOD1 promotor	SOD1, ubiquitin-positive and TDP-43-negative inclusion by 6 months	Motor impairment starting from 8 months	Denervation of muscle fiber. By end stage, spinal cord motor neurons are lost	Paralysis and death by 10 months	[87]
SOD1-G37R	SOD1 G37R	Transgene expressing human SOD1-G37R driven by human SOD1 promotor	SOD1 accumulate in axons	Progressive motor impairment.	NMJ denervation visible. Motor neurons in spinal cord loss by 5 months. No loss of upper motor neurons	Mice survive about 6-8 months	[88]
TARDBP (A315T)	TARDBP A315T	Transgene containing human TARDBP A315T with N terminal flag tag under mouse prion protein (PrP) promotor	ubiquitin-positive aggregates in motor neuron. No evidence for cytoplasmic TDP-43 inclusion	Impaired motor impairment	20% reduction in MNs from L3-L5 region. No neuromuscular deficits	Mice survive on average 153 days	[89, 90]
TDP-43 (A315T)	TARDBP A315T	Expression of Human TDP-43 (A315T) by endogenous human promotor	TDP-43 and ubiquitin-positive cytoplasmic aggregates by 10 months Increased C terminal fragment of TDP-43	Age-associated cognitive and motor deficits	No neuronal loss observed	N/A	[91]
TDP-43 (G348C)	TARDP G348C	Expression of Human TDP-43 (G348C) by endogenous human promotor	Cytoplasmic ubiquitin- and TDP-43-positive inclusion by 10 months	Age-associated cognitive and motor deficits	NMJ abnormality by 10 months No neuronal loss observed.	N/A	[91]

TDP-43 (M337V)	TARDBP M337V	Expression of Human TDP-43 (M337V) by mouse prion promoter	Ubiquitination and accumulation of phospho TDP-43 in cytoplasm and nucleus. Increase in lower molecular weight TDP-43 levels. Phospho tau detected	Body tremor and gait difficulty by 1 month	No neuronal loss	70% of Homozygotes die aroura 1 month	[92]
hTDP-43 ΔNLS		Conditional expression of hTDP-43 lacking Nuclear localization signal using Cank2a-tTA	Although TDP-43 is localized in cytoplasm, aggregate formation was rare. Downregulation of endogenous mouse TDP-43	Motor, cognitive and social deficits observed	Neuronal loss in dentate gyrus and cortical layers. No loss of lower motor neurons	N/A	[93]
FUS-R521C	FUS R521C	Expression of Flag tagged FUS R521C drive by hamster prion protein promotor	Cytoplasmic FUS inclusions are rare	Motor deficits	50% loss of neurons at the end stage	N/A	[94]
FUSΔNLS		Exclusion of exon 15 containing NLS sequence via cre-loxP system	Ubiquitin and p62 positive ΔNLS FUS inclusion present in Spinal cord neurons	Progressive motor impairments by 12 weeks	Neuronal loss in motor cortex by 1 year	50% mortality by 60 weeks	[95]
C9ORF72 (AAV)(G4C2)149	GGGGCC repeat	AAV mediated delivery of (GGGGCC) 149 repeat motif with 5′ and 3′ flanking regions under β-actin promotor into the lateral ventricles	Intranuclear RNA foci. Cytoplasmic inclusion of Dipeptide Repeat (DPRs)proteins. Accumulation of phosphorylated TDP-43	Motor and cognitive impairment by 3 months	Cortical neuron loss by 6 months	N/A	[96]
C9ORF72 (AAV)(G4C2)66	GGGGCC repeat	AAV mediated delivery of (GGGGCC) 66 repeat into the cerebral ventricles	Intranuclear RNA foci. DPR aggregates. Cytoplasmic aggregation of TDP-43 by 6 month	Subtle motor impairment and behavioral deficits	By 6 months, 17% fewer neurons in cortex and 11% fewer Purkinje cells	N/A	[97]
C9-BACexp	(GGGGCC)100-1000 repeats	BAC containing full-length human C9ORF72 with 100–1000 HRE repeats	RNA foci visible in neurons from 3 months. Soluble and insoluble poly (GP) DPR visible at 6 months. No TDP-43 aggregates	No behavioral abnormalities	No sign of neuronal loss	N/A	[98]

(continued)

Table 10.3 (continued)

Mouse model of ALS	Gene/mutation	Modification info	Neuropathology	Motor impairment	Muscle atrophy and neuron loss	Premature death	Ref.
C9-BAC500	(GGGGCC)500 repeats	BAC containing exon1 to exon6 of human C9ORF72 with 500 HRE repeats with 5′ upstream sequence	RNA foci visible in neurons from 3 months. Poly(GP) visible in neurons. No TDP-43 aggregates	No behavioral abnormalities	Neuronal loss	No alteration in survival	[99]
AAV-GFP-(GA)50	Poly(GA) 50 repeats	AAV vector encoding 50 repeats of GA with GFP tag	Presence of ubiquitin-positive poly(GA) aggregates at 4–6 weeks. Rare TDP-43 inclusion	Visible motor deficits by 6 months	Visible neuronal loss at 6 months	N/A	[100]
AAV-GFP-(GR)100	Poly(GR) 100 repeats	AAV vector encoding 100 repeats of GR with GFP tag	Diffused cytoplasmic poly(GR). Rare TDP-43 inclusion.	Visible motor deficits by 3 months	Visible neuronal loss in cortex by 6 months	N/A	[101]

than 50 different mutations have been identified in the TARDBP gene and the majority of ALS-linked mutations are found in the glycine-rich region of TDP-43 protein [108]. Numerous transgenic mouse models expressing human TDP-43 have been generated under various promoters. The TARDBP (A53T) (hybrid) or Baloh's TDP-43 transgenic mouse model was one of the first TDP-43 mouse models expressing human TDP-43 with A315T mutation under prion promoter [89]. These mice developed several features of the ALS pathology, including motor impairment, loss of motor neurons, presence of ubiquitin-positive inclusion and premature death, but they lacked cytoplasmic TDP-43 aggregates [89]. However, another transgenic mouse model with human TDP-43 A315 transgene under endogenous human promoter developed the pathological hallmark relevant to ALS/FTD, including cytoplasmic TDP-43 inclusion, but did not develop paralysis [91]. Interestingly, overexpression of wild-type human TDP-43 in mice is alone enough to induce ALS phenotype independent of mutations [109].

Fused in Sarcoma (FUS) is a 526 amino acid long protein ubiquitously expressed with predominant nuclear localization. It is involved in various cellular processes including cell proliferation, DNA repair, transcription regulation, and RNA procession [110]. To date, more than 50 different mutations in FUS have been identified in the ALS family, the majority of which are missense mutations clustered in C-terminal nuclear localization signals (NLS). These mutations have been shown to disrupt transportin-mediated nuclear import of FUS, leading to cytoplasmic FUS accumulation [111, 112]. The degree of disruption of NLS and the severity of the ALS phenotype are directly related to FUS R521C mutation. This results in moderate mislocalization of FUS to the cytoplasm, causing classical adult-onset ALS. In contrast, FUS P525L results in a more pronounced mislocalization and is therefore linked to juvenile-onset ALS [81, 113]. There are different transgenic mouse models expressing either wild-type FUS or FUS with R521C or P525L mutations or lacking NLS. The conditional knockout of FUS has no effect on neuronal survival and function; however, overexpression of either wild-type FUS or FUS with mutations is toxic and shows reduced lifespan in rodents [114, 115]. Interestingly, the conditional knock-in mouse model expressing FUS P525L mutant showed mislocalization of FUS to the cytoplasm, but does not form discrete aggregates. Nevertheless, they show progressive loss of motor neurons along with functional abnormalities at the neuromuscular junction [114].

Although as early as 2006, the gene in chromosome 9 was linked to families with ALS and FTD, it was only in 2011 that the genetic defect was identified as (GGGGCC) hexanucleotide repeat expansion (HRE) in the first intron of the C9ORF72 gene [83, 84, 116, 117]. The pathological hexanucleotide expansion appeared with a frequency of up to 29% of all FTD cases, 50% of all ALS cases, and 88% of all FTD/ALS cases. Currently, three different non-mutually exclusive pathomechanisms have been proposed to induce neurodegeneration through C9ORF72 HRE mutation. Firstly, reduced C9ORF72 protein expression due to the repeat expansion may lead to loss of function mechanism. The second hypothesis is the repeat-associated RNA toxicity whereby GGGGCC repeats aggregates into RNA foci, sequestering various RNA-binding proteins. The third hypothesis is the formation of toxic dipeptide repeat proteins (DPRs) produced due to repeat associated non-ATG translation of hexanucleotide repeat expansion [118]. Since the discovery of C9ORF72 mutations, significant efforts have been made to generate mouse models based on HRE and its different pathological hypotheses.

Transgenic mice with Knockout of 3110043O21 Rik, a murine homolog of human C9ORF72, do not develop motor neuron phenotype; however, they show defective endosomal trafficking and an increase in inflammatory markers suggesting that a loss of function of C9ORF72 alone is not sufficient to develop ALS phenotype [119]. Recently, C9ORF72 transgenic mice expressing HRE have been generated either by means of somatic brain transgenesis by adeno-associated virus (AAV) or under a human C9ORF72 promoter in bacterial artificial chromosome (BAC) [96, 97, 99]. The mice injected with AAV construct, with either 66 or 149 repeats, developed several pathological features of ALS/FTD including the formation of RNA foci, aggregation of different DPRs and cytoplasmic inclusions positive for phosphorylated TDP-43. The mice also displayed neuronal loss and developed behavioral and motor deficits [96, 97]. However, C9-BAC mice expressing the human C9ROF72 gene with 100–1000 repeats failed to show motor impairment and neuronal loss, despite recapitulating pathological hallmarks of C9ORF72 ALS, including the formation of RNA foci and the presence of different dipeptide repeat proteins. In addition, they also lacked phosphorylated TDP-43 and did not show evidence of increased microgliosis and astrogliosis [98, 99].

Moreover, AAV viruses encoding different DPRs have also been expressed in mice to study the effect of individual DPR, independent of repeat expansions [100, 101]. Mice expressing 100 repeats of GR showed progressive neuronal loss and showed motor and cognitive impairments. However, TDP-43 inclusions were rare and were observed only after 6 months [101]. The expression of 50 repeats of GA in mice resulted in the formation of ubiquitin-positive GA aggregates, progressive loss of neurons and motor and cognitive impairments [100].

10.2   Use of Non-rodent Model Organisms in Neurodegenerative Disease

Transgenic mice are widely used in studying neurodegenerative disorders and have been valuable in elucidating the pathomechanism of different diseases. Other than mice, different model organisms, like drosophila [120], *C. elegans* [121], zebrafish [122], and yeast [123] are also widely used to study various neurodegenerative diseases. There is high genomic conservation between species, and most of the disease-causing mutations are present in highly conserved domains, making each model organism a valuable tool to understand different human diseases. These small organism models of neurodegenerative diseases are able to recapitulate several aspects of human neurodegenerative diseases including protein aggregation, locomotor dysfunction, and reduced life span [120–123].

	Yeast	C. elegans	Drosophila	Zebrafish	Mouse	Non-human primates	iPSC-derived neurons
Life span	20 days	2/3 weeks	4/6 weeks	2-3 years	1-3 years	20 years	N/A
Gene homology for human disease	23%	65%	77%	84%	>90%	>95%	100%
Ease of genetic manipulation	★★★★★	★★★★★	★★★★★	★★★★☆	★★★☆☆	★☆☆☆☆	★★★★☆
Safety and ethics considerations	★★★★★	★★★★★	★★★★★	★★★★☆	★★☆☆☆	★☆☆☆☆	★★★★☆
Genome wide screening	★★★★★	★★★★☆	★★★★☆	★★★☆☆	☆☆☆☆☆	☆☆☆☆☆	★★★☆☆
Relevance to human physiology	★☆☆☆☆	★★☆☆☆	★★☆☆☆	★★★☆☆	★★★★☆	★★★★★	★★★★★
Relative cost	$	$	$	$$	$$$$	$$$$$	$$$$

Fig. 10.6 Comparison of commonly used model organisms in neurodegenerative diseases based on different features. Number of stars represents the degree of difficulty. Lower number of blue stars indicate a greater degree of difficulty

Each of these model organisms has its unique advantages and strengths (Fig. 10.6). A simple organism with a short generation time offers speed and easy genetic manipulation when compared with a higher organism. One of the powerful and widely used applications of smaller organism models is the identification of genetic modifiers of disease phenotype. The availability of overexpressing and knockout lines of different genes in these small organism models facilitates the use of unbiased neurodegenerative disease modifier screens. In addition, these small organisms can be used for high throughput screens of small molecules that can suppress the disease phenotype and reveal potential therapeutic targets. Despite the usefulness of smaller model organisms for neurodegenerative research, there are physiological limitations to the simulation of human disease. Hence, it is necessary to validate in mammalian models the knowledge gained through the use of small organisms.

10.3   Use of iPSCs to Model Neurodegenerative Diseases

Although there are various mouse models developed for different neurodegenerative diseases, none of them is a perfect model that recapitulates all the human disease pathobiology. For example, 5XFAD mice show the formation of Aβ plaque along with neuronal loss, but do not form neurofibrillary tangles [25, 26]. Likewise, C9 BAC mice show ALS pathology such as the formation of RNA foci and the presence of DPR aggregates. However, they do not result in neuronal death [98, 99]. Nevertheless, these models are still invaluable as they display certain disease features and have been useful in understanding pathomechanisms. As different mouse models only represent a part of the disease process, we must first understand the strength and limitations of the mouse models we are using for research and validate them in various other model systems.

Induced pluripotent stem cells (iPSCs) are cells that are reprogrammed from somatic cells by introducing four essential pluripotency factors, c-Myc, Klf4, Sox2, and Oct3/4, collectively termed as Yamanaka factors [124, 125]. The pluripotent cells can then be differentiated into desired cell types including astrocytes [126], oligodendrocytes [127], cortical neurons [128], or motor neurons [129] using modified protocols. This model system has several advantages over other model systems to study human neurodegenerative diseases, specifically iPSC technology models neurodegenerative diseases without the need for overexpressing the mutant gene. Additionally, the cells carry the complete genetic information of the donor patients, which makes it easy to capture the sporadic cases for which there is no model organism to date. Likewise, correcting the disease-causing mutation by using gene editing technologies will create the isogenic control line, which makes it a perfect control to identify both disease modifiers as well as disease mechanisms. Lastly, with the advances in the development of differentiation protocols, iPSCs can be differentiated into various neuronal subtypes as well as other cell types, enabling the investigation of mechanisms behind the cell type-specific vulnerability, as it is easy to capture proteomic or transcriptomic changes from a single cell type.

To date, many different neurodegenerative diseases have been studied using disease-specific neuronal cell types derived from patients' iPSCs (Fig. 10.7). iPSC-derived neurons generated from familial AD patients presented enhanced production of Aβ42, Aβ40, and Aβ42/40 ratio [130–132]. In addition to altered APP processing, neurons derived from both familial AD and sporadic AD cases, iPSCs also exhibited an increase in the levels of total tau and phosphorylated tau [131–133]. APOE4 allele is a major genetic risk factor for sporadic AD. Neurons derived from iPSCs carrying ApoE4/4 allele showed elevated levels of phosphorylated tau independent of the observed altered APP processing in neurons. Likewise, glia cells derived from ApoE4/4 iPSCs failed to clear extracellular Aβ proteins [134, 135]. Early pathogenic events like ER stress and activation of unfolded protein response (UPR) were also found to be upregulated in iPSC-derived neurons from familial AD and sporadic AD cases. In addition, altered cellular processes linked to the AD pathomechanism, involving proteasomal and lysosomal function, increased ROS production, reduced autophagy, and mitochondrial dysfunction, and were observed in patient iPSC-derived neurons [130, 136, 137].

Similarly, different studies have utilized reprogramming technologies to generate iPSCs from both sporadic PD and familial PD, and have used various differentiation protocols to generate midbrain dopaminergic neurons [138–142]. Dopaminergic neurons differentiated from iPSCs generated from the patient harboring A53T mutation displayed neuropathological features including phosphorylation of α-synuclein on serine 129 and aggregation that closely resembled the human PD cases [143]. Similarly, dopaminergic

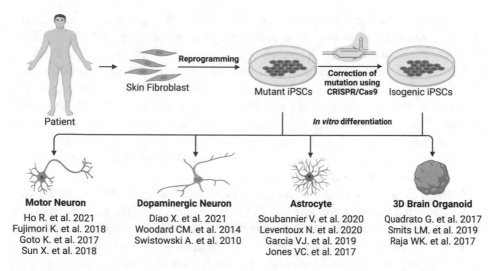

Fig. 10.7 Scheme depicting the generation of different cell types from patient-derived iPSC and gene editing methods to generate isogenic iPSCs

neurons differentiated from iPSCs derived from patients harboring mutations in Parkin and PINK1 genes also recapitulated PD phenotypes, including α-synuclein accumulation, mitochondrial dysfunction, and neuronal death [144]. Mutations in the GBA1 gene encoding for glucocerebrosidase is the strongest genetic risk factor for PD, and dopaminergic neurons generated from iPSC-carrying GBA1 mutation showed reduced glucocerebrosidase protein level and activity. Furthermore, defects in autophagy and lysosomal pathways resulted in enhanced α-synuclein aggregation, while isogenic gene-corrected controls did not show the pathological phenotypes [145].

In recent years, several studies have used iPSC-derived motor neurons from different ALS patients to model ALS pathology. These models recapitulate the neuropathology as seen in ALS patients, like SOD1 aggregates, TDP-43 aggregation, and the presence of RNA foci. Different RNA-binding proteins, including hnRNPA1 and Pur-α were found to be sequestered by RNA foci in neurons derived from C9ORF72 patients' iPSCs. Notably, treatment of neurons with antisense oligonucleotide (ASO) against HRE resulted in the dissolution of RNA foci and mitigated the C9ORF72 HRE-mediated toxicity [146]. iPSCs-derived neurons also reproduce different pathomechanisms related to ALS including ER stress, mitochondrial dysfunction, glutamate excitotoxicity, proteasomal dysfunction, etc.

2D neuronal cultures from patient-derived iPSCs have shown a resemblance to human pathology, and reproduce essential features of the pathology. Different studies have also used co-culture of neurons and astrocytes to study the influence of cell-autonomous and non-cell autonomous contributions to disease progression and cell death. Although iPSCs-derived neurons or astrocytes in 2D culture have yielded important findings due to their ability to capture the patient's genetic information, they fail to recapitulate the complexity of the human brain. The development of the iPSC technology has led to the generation of 3D brain organoids, which are able to model many complex features of the human brain. 3D brain organoids can be developed into distinct regions of the brain, such as forebrain, mid-brain, hippocampus, having an organized structure [147]. 3D brain organoids derived from iPSCs generated from familial AD patients have been able to recapitulate AD-like pathologies, including amyloid aggregation and hyperphosphorylated TAU in an age-dependent manner [148]. Brain organoids have also been used as a tool to access the effect of different pharmacological compounds on disease progression, for example, the treatment of AD patient-derived organoids with secretase inhibitors was seen to result in the reduction of amyloid and tau pathology [148].

Although iPSCs have been a valuable tool in capturing the patients' genetic information, it does not maintain the ages and epigenetic signatures of the patients. However, recently, motor neurons generated from the direct reprogramming of patients' fibroblast were able to maintain the aging hallmarks of patients, including DNA damage. This resulted in a more apt model to study late-onset neurodegenerative diseases [149]. Nevertheless, the low conversion rate of fibroblasts to neurons, as well as the limited number of neuronal subtypes generated, remains the main disadvantage of this model system.

Question

Describe the etiology and mutations associated with amyotrophic lateral sclerosis. Mention two model systems which are classically employed to study disease.

Take-Home Message

Although different mouse models and non-mammalian models have been valuable tools for understanding the pathomechanisms of different neurodegenerative diseases, there is limited success in translating the findings to useful clinical intervention. One of the main hindrances is that we can only model familial forms of the diseases, and the latter account for only a small fraction of the total cases. Even familial cases are linked to mutations in different genes resulting in varying pathology. However, all models have their own strengths and weaknesses. Hence, more attention should be paid to choosing the correct model based on the relevance of the question that is being asked. Humans and rodents differ in many ways; hence, we must be cautious while interpreting different behavioral deficits. As mouse models cannot reflect the genetic diversity of human population, studies should also focus on validating the findings in patient-derived materials, such as iPSC-derived neurons or post-mortem tissues.

References

1. Soto C. Unfolding the role of protein misfolding in neurodegenerative diseases. Nat Rev Neurosci. 2003;4(1):49–60.
2. Ingelsson M. Alpha-synuclein oligomers-neurotoxic molecules in Parkinson's disease and other Lewy body disorders. Front Neurosci. 2016;10:1–10.
3. Perl DP. Neuropathology of Alzheimer's disease. Mt Sinai J Med. 2010;77(1):32–42.
4. Mori F, Tanji K, Zhang H-X, Nishihira Y, Tan C-F, Takahashi H, et al. Maturation process of TDP-43-positive neuronal cytoplasmic inclusions in amyotrophic lateral sclerosis with and without dementia. Acta Neuropathol. 2008;116(2):193–203.
5. Vatsavayai SC, Yoon SJ, Gardner RC, Gendron TF, Vargas JNS, Trujillo A, et al. Timing and significance of pathological features in C9orf72 expansion-associated frontotemporal dementia. Brain. 2016;139(Pt 12):3202–16.
6. Paré B, Lehmann M, Beaudin M, Nordström U, Saikali S, Julien JP, et al. Misfolded SOD1 pathology in sporadic amyotrophic lateral sclerosis. Sci Rep. 2018;8(1):1–13.
7. Waldvogel HJ, Kim EH, Tippett LJ, Vonsattel J-PG, Faull RLM. The neuropathology of Huntington's disease. In: Nguyen HHP, Cenci MA, editors. Behavioral neurobiology of Huntington's disease and Parkinson's disease. Berlin: Springer; 2015. p. 33–80.
8. Gan L, Cookson MR, Petrucelli L, La Spada AR. Converging pathways in neurodegeneration, from genetics to mechanisms. Nat Neurosci. 2018;21(10):1300–9.
9. Dawson TM, Golde TE, Lagier-Tourenne C. Animal models of neurodegenerative diseases. Nat Neurosci. 2018;21(10):1370–9.
10. Crous-Bou M, Minguillón C, Gramunt N, Molinuevo JL. Alzheimer's disease prevention: from risk factors to early intervention. Alzheimers Res Ther. 2017;9(1):71.
11. 2020 Alzheimer's disease facts and figures. Alzheimers Dement. 2020;16(3):391–460.
12. Long JM, Holtzman DM. Alzheimer disease: an update on pathobiology and treatment strategies. Cell. 2019;179(2):312–39.
13. Neve RL, McPhie DL, Chen Y. Alzheimer's disease: a dysfunction of the amyloid precursor protein(1). Brain Res. 2000;886(1–2):54–66.
14. Nelson PT, Alafuzoff I, Bigio EH, Bouras C, Braak H, Cairns NJ, et al. Correlation of Alzheimer disease neuropathologic changes with cognitive status: a review of the literature. J Neuropathol Exp Neurol. 2012;71(5):362–81.
15. Hardy JA, Higgins GA. Alzheimer's disease: the amyloid cascade hypothesis. Science. 1992;256(5054):184–5.
16. Hanseeuw BJ, Betensky RA, Jacobs HIL, Schultz AP, Sepulcre J, Becker JA, et al. Association of amyloid and tau with cognition in preclinical Alzheimer disease: a longitudinal study. JAMA Neurol. 2019;76(8):915–24.
17. Research models: Alzheimer's disease [Internet]. Available from: https://www.alzforum.org/research-models/alzheimers-disease
18. Quon D, Wang Y, Catalano R, Scardina JM, Murakami K, Cordell B. Formation of β-amyloid protein deposits in brains of transgenic mice. Nature. 1991;352(6332):239–41.
19. Higgins LS, Holtzman DM, Rabin J, Mobley WC, Cordell B. Transgenic mouse brain histopathology resembles early Alzheimer's disease. Ann Neurol. 1994;35(5):598–607.
20. Games D, Adams D, Alessandrini R, Barbour R, Borthelette P, Blackwell C, et al. Alzheimer-type neuropathology in transgenic mice overexpressing V717F β-amyloid precursor protein. Nature. 1995;373(6514):523–7.
21. Karen H, Paul C, Steven N, Chris E, Yasuo H, Steven Y, et al. Correlative memory deficits, Aβ elevation, and amyloid plaques in transgenic mice. Science. 1996;274(5284):99–103.

22. Sturchler-Pierrat C, Abramowski D, Duke M, Wiederhold K-H, Mistl C, Rothacher S, et al. Two amyloid precursor protein transgenic mouse models with Alzheimer disease-like pathology. Proc Natl Acad Sci USA. 1997;94(24):13287–92.
23. Bi C, Bi S, Li B. Processing of mutant β-amyloid precursor protein and the clinicopathological features of familial Alzheimer's disease. Aging Dis. 2019;10(2):383–403.
24. Duff K, Eckman C, Zehr C, Yu X, Prada C-M, Perez-tur J, et al. Increased amyloid-β42(43) in brains of mice expressing mutant presenilin 1. Nature. 1996;383(6602):710–3.
25. Oakley H, Cole SL, Logan S, Maus E, Shao P, Craft J, et al. Intraneuronal β-amyloid aggregates, neurodegeneration, and neuron loss in transgenic mice with five familial Alzheimer's disease mutations: potential factors in amyloid plaque formation. J Neurosci. 2006;26(40):10129–40.
26. Jawhar S, Trawicka A, Jenneckens C, Bayer TA, Wirths O. Motor deficits, neuron loss, and reduced anxiety coinciding with axonal degeneration and intraneuronal Aβ aggregation in the 5XFAD mouse model of Alzheimer's disease. Neurobiol Aging. 2012;33(1):196.e29–40.
27. Rockenstein EM, McConlogue L, Tan H, Power M, Masliah E, Mucke L. Levels and alternative splicing of amyloid β protein precursor (APP) transcripts in brains of APP transgenic mice and humans with Alzheimer's disease. J Biol Chem. 1995;270(47):28257–67.
28. Oddo S, Caccamo A, Shepherd JD, Murphy MP, Golde TE, Kayed R, et al. Triple-transgenic model of Alzheimer's disease with plaques and tangles: intracellular Aβ and synaptic dysfunction. Neuron. 2003;39(3):409–21.
29. Lewis J, McGowan E, Rockwood J, Melrose H, Nacharaju P, Van Slegtenhorst M, et al. Neurofibrillary tangles, amyotrophy and progressive motor disturbance in mice expressing mutant (P301L) tau protein. Nat Genet. 2000;25(4):402–5.
30. Jang H, Ryu JH, Shin KM, Seo N-Y, Kim GH, Huh YH, et al. Gait ignition failure in JNPL3 human tau-mutant mice. Exp Neurobiol. 2019;28(3):404–13.
31. Billings LM, Oddo S, Green KN, McGaugh JL, LaFerla FM. Intraneuronal Aβ causes the onset of early Alzheimer's disease-related cognitive deficits in transgenic mice. Neuron. 2005;45(5):675–88.
32. Jankowsky JL, Younkin LH, Gonzales V, Fadale DJ, Slunt HH, Lester HA, et al. Rodent Aβ modulates the solubility and distribution of amyloid deposits in transgenic mice. J Biol Chem. 2007;282(31):22707–20.
33. Hernández F, Merchán-Rubira J, Vallés-Saiz L, Rodríguez-Matellán A, Avila J. Differences between human and murine tau at the N-terminal end. Front Aging Neurosci. 2020;12:11.
34. Del Rey NL-G, Quiroga-Varela A, Garbayo E, Carballo-Carbajal I, Fernández-Santiago R, Monje MHG, et al. Advances in Parkinson's disease: 200 years later. Front Neuroanat. 2018;12:113.
35. Spillantini MG, Goedert M. The α-synucleinopathies: Parkinson's Disease, Dementia with Lewy bodies, and multiple system atrophy. Ann N Y Acad Sci. 2000;920(1):16–27.
36. Fares MB, Jagannath S, Lashuel HA. Reverse engineering Lewy bodies: how far have we come and how far can we go? Nat Rev Neurosci. 2021;22(2):111–31.
37. Braak H, Rüb U, Gai WP, Del Tredici K. Idiopathic Parkinson's disease: possible routes by which vulnerable neuronal types may be subject to neuroinvasion by an unknown pathogen. J Neural Transm. 2003;110(5):517–36.
38. Kordower JH, Chu Y, Hauser RA, Freeman TB, Olanow CW. Lewy body-like pathology in long-term embryonic nigral transplants in Parkinson's disease. Nat Med. 2008;14(5):504–6.
39. Li J-Y, Englund E, Holton JL, Soulet D, Hagell P, Lees AJ, et al. Lewy bodies in grafted neurons in subjects with Parkinson's disease suggest host-to-graft disease propagation. Nat Med. 2008;14(5):501–3.

40. Luk KC, Song C, O'Brien P, Stieber A, Branch JR, Brunden KR, et al. Exogenous α-synuclein fibrils seed the formation of Lewy body-like intracellular inclusions in cultured cells. Proc Natl Acad Sci USA. 2009;106(47):20051–6.

41. Mahul-Mellier A-LL, Burtscher J, Maharjan N, Weerens L, Croisier M, Kuttler F, et al. The process of Lewy body formation, rather than simply α-synuclein fibrillization, is one of the major drivers of neurodegeneration. Proc Natl Acad Sci U S A. 2020;117(9):4971–82.

42. Luk KC, Kehm V, Carroll J, Zhang B, O'Brien P, Trojanowski JQ, et al. Pathological α-synuclein transmission initiates Parkinson-like neurodegeneration in nontransgenic mice. Science. 2012;338(6109):949–53.

43. Masuda-Suzukake M, Nonaka T, Hosokawa M, Oikawa T, Arai T, Akiyama H, et al. Prion-like spreading of pathological α-synuclein in brain. Brain. 2013;136(4):1128–38.

44. Okuzumi A, Kurosawa M, Hatano T, Takanashi M, Nojiri S, Fukuhara T, et al. Rapid dissemination of alpha-synuclein seeds through neural circuits in an in-vivo prion-like seeding experiment. Acta Neuropathol Commun. 2018;6(1):96.

45. Thakur P, Breger LS, Lundblad M, Wan OW, Mattsson B, Luk KC, et al. Modeling Parkinson's disease pathology by combination of fibril seeds and α-synuclein overexpression in the rat brain. Proc Natl Acad Sci USA. 2017;114(39):E8284–93.

46. Peelaerts W, Bousset L, Van der Perren A, Moskalyuk A, Pulizzi R, Giugliano M, et al. α-Synuclein strains cause distinct synucleinopathies after local and systemic administration. Nature. 2015;522(7556):340–4.

47. Chung HK, Ho H-A, Pérez-Acuña D, Lee S-J. Modeling α-synuclein propagation with preformed fibril injections. J Mov Disord. 2019;12(3):139–51.

48. Kim WS, Kågedal K, Halliday GM. Alpha-synuclein biology in Lewy body diseases. Alzheimers Res Ther. 2014;6(5):73.

49. Lee MK, Stirling W, Xu Y, Xu X, Qui D, Mandir AS, et al. Human α-synuclein-harboring familial Parkinson's disease-linked Ala-53 → Thr mutation causes neurodegenerative disease with α-synuclein aggregation in transgenic mice. Proc Natl Acad Sci USA. 2002;99(13):8968–73.

50. Kuo Y-M, Li Z, Jiao Y, Gaborit N, Pani AK, Orrison BM, et al. Extensive enteric nervous system abnormalities in mice transgenic for artificial chromosomes containing Parkinson disease-associated α-synuclein gene mutations precede central nervous system changes. Hum Mol Genet. 2010;19(9):1633–50.

51. Richfield EK, Thiruchelvam MJ, Cory-Slechta DA, Wuertzer C, Gainetdinov RR, Caron MG, et al. Behavioral and neurochemical effects of wild-type and mutated human α-synuclein in transgenic mice. Exp Neurol. 2002;175(1):35–48.

52. Tofaris GK, Garcia Reitböck P, Humby T, Lambourne SL, O'Connell M, Ghetti B, et al. Pathological changes in dopaminergic nerve cells of the substantia nigra and olfactory bulb in mice transgenic for truncated human alpha-synuclein(1-120): implications for Lewy body disorders. J Neurosci. 2006;26(15):3942–50.

53. Rockenstein E, Mallory M, Hashimoto M, Song D, Shults CW, Lang I, et al. Differential neuropathological alterations in transgenic mice expressing α-synuclein from the platelet-derived growth factor and Thy-1 promoters. J Neurosci Res. 2002;68(5):568–78.

54. Cabin DE, Shimazu K, Murphy D, Cole NB, Gottschalk W, McIlwain KL, et al. Synaptic vesicle depletion correlates with attenuated synaptic responses to prolonged repetitive stimulation in mice lacking α-synuclein. J Neurosci. 2002;22(20):8797–807.

55. Lu X-H, Fleming SM, Meurers B, Ackerson LC, Mortazavi F, Lo V, et al. Bacterial artificial chromosome transgenic mice expressing a truncated mutant parkin exhibit age-dependent hypokinetic motor deficits, dopaminergic neuron degeneration, and accumulation of proteinase K-resistant α-synuclein. J Neurosci. 2009;29(7):1962–76.

56. Kitada T, Pisani A, Porter DR, Yamaguchi H, Tscherter A, Martella G, et al. Impaired dopamine release and synaptic plasticity in the striatum of PINK1-deficient mice. Proc Natl Acad Sci USA. 2007;104(27):11441–6.

57. Gispert S, Ricciardi F, Kurz A, Azizov M, Hoepken H-H, Becker D, et al. Parkinson phenotype in aged PINK1-deficient mice is accompanied by progressive mitochondrial dysfunction in absence of neurodegeneration. PLoS One. 2009;4(6):e5777.

58. Tsika E, Kannan M, Foo CS-Y, Dikeman D, Glauser L, Gellhaar S, et al. Conditional expression of Parkinson's disease-related R1441C LRRK2 in midbrain dopaminergic neurons of mice causes nuclear abnormalities without neurodegeneration. Neurobiol Dis. 2014;71:345–58.

59. Ramonet D, Daher JPL, Lin BM, Stafa K, Kim J, Banerjee R, et al. Dopaminergic neuronal loss, reduced neurite complexity and autophagic abnormalities in transgenic mice expressing G2019S mutant LRRK2. PLoS One. 2011;6(4):e18568.

60. Li X, Patel JC, Wang J, Avshalumov MV, Nicholson C, Buxbaum JD, et al. Enhanced striatal dopamine transmission and motor performance with LRRK2 overexpression in mice is eliminated by familial Parkinson's disease mutation G2019S. J Neurosci. 2010;30(5):1788–97.

61. Xiong Y, Neifert S, Karuppagounder SS, Liu Q, Stankowski JN, Lee BD, et al. Robust kinase- and age-dependent dopaminergic and norepinephrine neurodegeneration in LRRK2 G2019S transgenic mice. Proc Natl Acad Sci USA. 2018;115(7):1635–40.

62. Schneider JS, Anderson DW, Decamp E. 1-methyl-4-phenyl-1,2,3,6-tetrahydropyridine-induced Mammalian Models of Parkinson's disease. In: Nass R, Przedborski S, editors. Parkinson's disease. San Diego: Elsevier; 2008. p. 87–103.

63. Hernandez-Baltazar D, Zavala-Flores LM, Villanueva-Olivo A. The 6-hydroxydopamine model and parkinsonian pathophysiology: novel findings in an older model. Neurologia (English Ed). 2017;32(8):533–9.

64. Duty S, Jenner P. Animal models of Parkinson's disease: a source of novel treatments and clues to the cause of the disease. Br J Pharmacol. 2011;164(4):1357–91.

65. Nieto M, Gil-Bea FJ, Dalfó E, Cuadrado M, Cabodevilla F, Sánchez B, et al. Increased sensitivity to MPTP in human α-synuclein A30P transgenic mice. Neurobiol Aging. 2006;27(6):848–56.

66. Kiernan MC, Vucic S, Cheah BC, Turner MR, Eisen A, Hardiman O, et al. Amyotrophic lateral sclerosis. Lancet. 2011;377(9769):942–55.

67. Ticozzi N, Tiloca C, Morelli C, Colombrita C, Poletti B, Doretti A, et al. Genetics of familial Amyotrophic lateral sclerosis. Arch Ital Biol. 2011;149(1):65–82.

68. Sabatelli M, Conte a, Zollino M. Clinical and genetic heterogeneity of amyotrophic lateral sclerosis. Clin Genet. 2013;83(5):408–16.

69. Neumann M, Sampathu DM, Kwong LK, Truax AC, Micsenyi MC, Chou TT, et al. Ubiquitinated TDP-43 in frontotemporal lobar degeneration and amyotrophic lateral sclerosis. Science. 2006;314(5796):130–3.

70. Borchelt DR, Lee MK, Slunt HS, Guarnieri M, Xu ZS, Wong PC, et al. Superoxide dismutase 1 with mutations linked to familial amyotrophic lateral sclerosis possesses significant activity. Proc Natl Acad Sci U S A. 1994;91(17):8292–6.

71. Bunton-Stasyshyn RKA, Saccon RA, Fratta P, Fisher EMC. SOD1 function and its implications for amyotrophic lateral sclerosis pathology: new and renascent themes. Neuroscientist. 2015;21(5):519–29.

72. Mackenzie IR, a. The neuropathology of FTD associated With ALS. Alzheimer Dis Assoc Disord. 2007;21(4):S44–9.

73. Arai T, Hasegawa M, Akiyama H, Ikeda K, Nonaka T, Mori H, et al. TDP-43 is a component of ubiquitin-positive tau-negative inclusions in frontotemporal lobar degeneration and amyotrophic lateral sclerosis. Biochem Biophys Res Commun. 2006;351(3):602–11.

74. Cairns NJ, Neumann M, Bigio EH, Holm IE, Troost D, Hatanpaa KJ, et al. TDP-43 in familial and sporadic frontotemporal lobar degeneration with ubiquitin inclusions. Am J Pathol. 2007;171(1):227–40.

75. Neumann M, Roeber S, Kretzschmar H, a, Rademakers R, Baker M, Mackenzie IR a. Abundant FUS-immunoreactive pathology in neuronal intermediate filament inclusion disease. Acta Neuropathol. 2009;118(5):605–16.

76. Lillo P, Hodges JR. Frontotemporal dementia and motor neurone disease: overlapping clinic-pathological disorders. J Clin Neurosci. 2009;16(9):1131–5.

77. Lillo P, Savage S, Mioshi E, Kiernan MC, Hodges JR. Amyotrophic lateral sclerosis and frontotemporal dementia: a behavioural and cognitive continuum. Amyotroph Lateral Scler. 2012 Jan;13(1):102–9.

78. Phukan J, Elamin M, Bede P, Jordan N, Gallagher L, Byrne S, et al. The syndrome of cognitive impairment in amyotrophic lateral sclerosis: a population-based study. J Neurol Neurosurg Psychiatry. 2012;83(1):102–8.

79. Benajiba L, Le BI, Camuzat A, Lacoste M, Thomas-Anterion C, Couratier P, et al. TARDBP mutations in motoneuron disease with frontotemporal lobar degeneration. Ann Neurol. 2009;65(4):470–4.

80. Sreedharan J, Blair IP, Tripathi VB, Hu X, Vance C, Rogelj B, et al. TDP-43 mutations in familial and sporadic amyotrophic lateral sclerosis. Science. 2008;319(5870):1668–72.

81. Kwiatkowski TJ, Bosco DA, Leclerc AL, Tamrazian E, Vanderburg CR, Russ C, et al. Mutations in the FUS/TLS gene on chromosome 16 cause familial amyotrophic lateral sclerosis. Science. 2009;323(5918):1205–8.

82. Ferrari R, Kapogiannis D, Huey ED, Momeni P. FTD and ALS: a tale of two diseases. Curr Alzheimer Res. 2011;8(3):273–94.

83. Renton AEE, Majounie E, Waite A, Simón-Sánchez J, Rollinson S, Gibbs JRR, ct al. A hexanucleotide repeat expansion in C9ORF72 is the cause of chromosome 9p21-linked ALS-FTD. Neuron. 2011;72(2):257–68.

84. DeJesus-Hernandez M, Mackenzie IRR, Boeve BFF, Boxer ALL, Baker M, Rutherford NJJ, et al. Expanded GGGGCC hexanucleotide repeat in noncoding region of C9ORF72 causes chromosome 9p-linked FTD and ALS. Neuron. 2011;72(2):245–56.

85. Alsultan AA, Waller R, Heath PR, Kirby J. The genetics of amyotrophic lateral sclerosis: current insights. Degener Neurol Neuromuscul Dis. 2016;(6):49–64.

86. Gurney ME, Pu H, Chiu A, Dal Canto M, Polchow C, Alexander D, et al. Motor neuron degeneration in mice that express a human Cu,Zn superoxide dismutase mutation. Science. 1994;264(5166):1772–5.

87. Bruijn LI, Becher MW, Lee MK, Anderson KL, Jenkins NA, Copeland NG, et al. ALS-linked SOD1 mutant G85R mediates damage to astrocytes and promotes rapidly progressive disease with SOD1-containing inclusions. Neuron. 1997;18(2):327–38.

88. Wong PC, Pardo CA, Borchelt DR, Lee MK, Copeland NG, Jenkins NA, et al. An adverse property of a familial ALS-linked SOD1 mutation causes motor neuron disease characterized by vacuolar degeneration of mitochondria. Neuron. 1995;14(6):1105–16.

89. Wegorzewska I, Bell S, Cairns NJ, Miller TM, Baloh RH. TDP-43 mutant transgenic mice develop features of ALS and frontotemporal lobar degeneration. Proc Natl Acad Sci USA. 2009;106(44):18809–14.

90. Hatzipetros T, Bogdanik LP, Tassinari VR, Kidd JD, Moreno AJ, Davis C, et al. C57BL/6J congenic Prp-TDP43A315T mice develop progressive neurodegeneration in the myenteric plexus of the colon without exhibiting key features of ALS. Brain Res. 2014;1584:59–72.

91. Swarup V, Phaneuf D, Bareil C, Robertson J, Rouleau GA, Kriz J, et al. Pathological hallmarks of amyotrophic lateral sclerosis/frontotemporal lobar degeneration in transgenic mice produced with TDP-43 genomic fragments. Brain. 2011;134(9):2610–26.

92. Xu Y-F, Zhang Y-J, Lin W-L, Cao X, Stetler C, Dickson DW, et al. Expression of mutant TDP-43 induces neuronal dysfunction in transgenic mice. Mol Neurodegener. 2011;6(1):73.

93. Igaz LM, Kwong LK, Lee EB, Chen-Plotkin A, Swanson E, Unger T, et al. Dysregulation of the ALS-associated gene TDP-43 leads to neuronal death and degeneration in mice. J Clin Invest. 2011;121(2):726–38.

94. Qiu H, Lee S, Shang Y, Wang W-Y, Au KF, Kamiya S, et al. ALS-associated mutation FUS-R521C causes DNA damage and RNA splicing defects. J Clin Invest. 2014;124(3):981–99.

95. Scekic-Zahirovic J, Sendscheid O, El Oussini H, Jambeau M, Sun Y, Mersmann S, et al. Toxic gain of function from mutant FUS protein is crucial to trigger cell autonomous motor neuron loss. EMBO J. 2016;35(10):1077–97.

96. Chew J, Cook C, Gendron TF, Jansen-West K, del Rosso G, Daughrity LM, et al. Aberrant deposition of stress granule-resident proteins linked to C9orf72-associated TDP-43 proteinopathy. Mol Neurodegener. 2019;14(1):9.

97. Chew J, Gendron TF, Prudencio M, Sasaguri H, Zhang YJ, Castanedes-Casey M, et al. C9ORF72 repeat expansions in mice cause TDP-43 pathology, neuronal loss, and behavioral deficits. Science. 2015;348(6239):1151–4.

98. O'Rourke JG, Bogdanik L, Muhammad AKMG, Gendron TF, Kim KJ, Austin A, et al. C9orf72 BAC transgenic mice display typical pathologic features of ALS/FTD. Neuron. 2015;88(5):892–901.

99. Peters OM, Cabrera GT, Tran H, Gendron TF, McKeon JE, Metterville J, et al. Human C9ORF72 hexanucleotide expansion reproduces RNA foci and dipeptide repeat proteins but not neurodegeneration in BAC transgenic mice. Neuron. 2015;88(5):902–9.

100. Zhang Y, Gendron TF, Grima JC, Sasaguri H, Jansen-west K, Xu Y, et al. C9ORF72 poly (GA) aggregates sequester and impair HR23 and nucleocytoplasmic transport proteins. Nat Neurosci. 2016;19(5)

101. Zhang YJ, Gendron TF, Ebbert MTW, O'Raw AD, Yue M, Jansen-West K, et al. Poly(GR) impairs protein translation and stress granule dynamics in C9orf72-associated frontotemporal dementia and amyotrophic lateral sclerosis. Nat Med. 2018;24(8):1136–42.

102. Rosen DR. Mutations in Cu/Zn superoxide dismutase gene are associated with familial amyotrophic lateral sclerosis. Nature. 1993;362:38–62.

103. Reaume AG, Elliott JL, Hoffman EK, Kowall NW, Ferrante RJ, Siwek DF, et al. Motor neurons in Cu/Zn superoxide dismutase-deficient mice develop normally but exhibit enhanced cell death after axonal injury. Nat Genet. 1996;13(1):43–7.

104. Heiman-Patterson TD, Sher RB, Blankenhorn EA, Alexander G, Deitch JS, Kunst CB, et al. Effect of genetic background on phenotype variability in transgenic mouse models of amyotrophic lateral sclerosis: a window of opportunity in the search for genetic modifiers. Amyotroph Lateral Scler. 2011;12(2):79–86.

105. Robertson J, Sanelli T, Xiao S, Yang W, Horne P, Hammond R, et al. Lack of TDP-43 abnormalities in mutant SOD1 transgenic mice shows disparity with ALS. Neurosci Lett. 2007;420(2):128–32.

106. Yokoseki A, Shiga A, Tan C-F, Tagawa A, Kaneko H, Koyama A, et al. TDP-43 mutation in familial amyotrophic lateral sclerosis. Ann Neurol. 2008;63(4):538–42.

107. Prasad A, Bharathi V, Sivalingam V, Girdhar A, Patel BK. Molecular mechanisms of TDP-43 misfolding and pathology in amyotrophic lateral sclerosis. Front Mol Neurosci. 2019;12:25.

108. de Boer EMJ, Orie VK, Williams T, Baker MR, De Oliveira HM, Polvikoski T, et al. TDP-43 proteinopathies: a new wave of neurodegenerative diseases. J Neurol Neurosurg Psychiatry. 2021;92(1):86–95.

109. Xu Y-F, Gendron TF, Zhang Y-J, Lin W-L, D'Alton S, Sheng H, et al. Wild-type human TDP-43 expression causes TDP-43 phosphorylation, mitochondrial aggregation, motor deficits, and early mortality in transgenic mice. J Neurosci. 2010;30(32):10851–9.

110. Shang Y, Huang EJ. Mechanisms of FUS mutations in familial amyotrophic lateral sclerosis. Brain Res. 2016;1647:65–78.

111. Dormann D, Rodde R, Edbauer D, Bentmann E, Fischer I, Hruscha A, et al. ALS-associated fused in sarcoma (FUS) mutations disrupt Transportin-mediated nuclear import. EMBO J. 2010;29(16):2841–57.

112. Ito D, Suzuki N. Conjoint pathologic cascades mediated by ALS/FTLD-U linked RNA-binding proteins TDP-43 and FUS. Neurology. 2011;77(17):1636–43.

113. Conte A, Lattante S, Zollino M, Marangi G, Luigetti M, Del Grande A, et al. P525L FUS mutation is consistently associated with a severe form of juvenile amyotrophic lateral sclerosis. Neuromuscul Disord. 2012;22(1):73–5.

114. Sharma A, Lyashchenko AK, Lu L, Nasrabady SE, Elmaleh M, Mendelsohn M, et al. ALS-associated mutant FUS induces selective motor neuron degeneration through toxic gain of function. Nat Commun. 2016;7(1):10465.

115. Nolan M, Talbot K, Ansorge O. Pathogenesis of FUS-associated ALS and FTD: insights from rodent models. Acta Neuropathol Commun. 2016;4(1):99.

116. Morita M, Al-Chalabi A, Andersen PM, Hosler B, Sapp P, Englund E, et al. A locus on chromosome 9p confers susceptibility to ALS and frontotemporal dementia. Neurology. 2006;66(6):839–44.

117. Vance C, Al-Chalabi A, Ruddy D, Smith BN, Hu X, Sreedharan J, et al. Familial amyotrophic lateral sclerosis with frontotemporal dementia is linked to a locus on chromosome 9p13.2–21.3. Brain. 2006;129(4):868–76.

118. Schmitz A, Pinheiro Marques J, Oertig I, Maharjan N, Saxena S. Emerging perspectives on dipeptide repeat proteins in C9ORF72 ALS/FTD. Front Cell Neurosci. 2021;15:637548.

119. O'Rourke JG, Bogdanik L, Yáñez A, Lall D, Wolf AJ, Muhammad AKMG, et al. C9orf72 is required for proper macrophage and microglial function in mice. Science. 2016;351(6279):1324–9.

120. Marsh JL, Thompson LM. Drosophila in the study of neurodegenerative disease. Neuron. 2006;52(1):169–78.

121. Alexander AG, Marfil V, Li C. Use of Caenorhabditis elegans as a model to study Alzheimer's disease and other neurodegenerative diseases. Front Genet. 2014;5:1–21.

122. Wang X, Zhang J-B, He K-J, Wang F, Liu C-F. Advances of zebrafish in neurodegenerative disease: from models to drug discovery. Front Pharmacol. 2021;12:713963.

123. Miller-Fleming L, Giorgini F, Outeiro TF. Yeast as a model for studying human neurodegenerative disorders. Biotechnol J. 2008;3(3):325–38.

124. Park I-H, Zhao R, West JA, Yabuuchi A, Huo H, Ince TA, et al. Reprogramming of human somatic cells to pluripotency with defined factors. Nature. 2008;451(7175):141–6.

125. Junying Y, Vodyanik VM, Kim S-O, Jessica A-B, Frane JL, Shulan T, et al. Induced pluripotent stem cell lines derived from human somatic cells. Science. 2007;318(5858):1917–20.

126. Soubannier V, Maussion G, Chaineau M, Sigutova V, Rouleau G, Durcan TM, et al. Characterization of human iPSC-derived astrocytes with potential for disease modeling and drug discovery. Neurosci Lett. 2020;731:135028.

127. Ehrlich M, Mozafari S, Glatza M, Starost L, Velychko S, Hallmann A-L, et al. Rapid and efficient generation of oligodendrocytes from human induced pluripotent stem cells using transcription factors. Proc Natl Acad Sci USA. 2017;114(11):E2243–52.

128. Mehta SR, Tom CM, Wang Y, Bresee C, Rushton D, Mathkar PP, et al. Human Huntington's disease iPSC-Derived cortical neurons display altered transcriptomics, morphology, and maturation. Cell Rep. 2018;25(4):1081–1096.e6.

129. Bianchi F, Malboubi M, Li Y, George JH, Jerusalem A, Szele F, et al. Rapid and efficient differentiation of functional motor neurons from human iPSC for neural injury modelling. Stem Cell Res. 2018;32:126–34.

130. Kondo T, Asai M, Tsukita K, Kutoku Y, Ohsawa Y, Sunada Y, et al. Modeling Alzheimer's disease with iPSCs reveals stress phenotypes associated with intracellular Aβ and differential drug responsiveness. Cell Stem Cell. 2013;12(4):487–96.

131. Israel MA, Yuan SH, Bardy C, Reyna SM, Mu Y, Herrera C, et al. Probing sporadic and familial Alzheimer's disease using induced pluripotent stem cells. Nature. 2012;482(7384):216–20.

132. Muratore CR, Rice HC, Srikanth P, Callahan DG, Shin T, Benjamin LNP, et al. The familial Alzheimer's disease APPV717I mutation alters APP processing and Tau expression in iPSC-derived neurons. Hum Mol Genet. 2014;23(13):3523–36.

133. Yang J, Zhao H, Ma Y, Shi G, Song J, Tang Y, et al. Early pathogenic event of Alzheimer's disease documented in iPSCs from patients with PSEN1 mutations. Oncotarget. 2017;8(5):7900–13.

134. Lin Y-T, Seo J, Gao F, Feldman HM, Wen H-L, Penney J, et al. APOE4 causes widespread molecular and cellular alterations associated with Alzheimer's disease phenotypes in human iPSC-derived brain cell types. Neuron. 2018;98(6):1141–1154.e7.

135. Wang C, Najm R, Xu Q, Jeong D, Walker D, Balestra ME, et al. Gain of toxic apolipoprotein E4 effects in human iPSC-derived neurons is ameliorated by a small-molecule structure corrector. Nat Med. 2018;24(5):647–57.

136. Martín-Maestro P, Gargini R, Sproul AA, García E, Antón LC, Noggle S, et al. Mitophagy failure in fibroblasts and iPSC-derived neurons of Alzheimer's disease-associated presenilin 1 mutation. Front Mol Neurosci. 2017;10:291.

137. Birnbaum JH, Wanner D, Gietl AF, Saake A, Kündig TM, Hock C, et al. Oxidative stress and altered mitochondrial protein expression in the absence of amyloid-β and tau pathology in iPSC-derived neurons from sporadic Alzheimer's disease patients. Stem Cell Res. 2018;27:121–30.

138. Soldner F, Hockemeyer D, Beard C, Gao Q, Bell GW, Cook EG, et al. Parkinson's disease patient-derived induced pluripotent stem cells free of viral reprogramming factors. Cell. 2009;136(5):964–77.

139. Devine MJ, Ryten M, Vodicka P, Thomson AJ, Burdon T, Houlden H, et al. Parkinson's disease induced pluripotent stem cells with triplication of the α-synuclein locus. Nat Commun. 2011;2(1):440.

140. Woodard CM, Campos BA, Kuo S-H, Nirenberg MJ, Nestor MW, Zimmer M, et al. iPSC-derived dopamine neurons reveal differences between monozygotic twins discordant for Parkinson's disease. Cell Rep. 2014;9(4):1173–82.

141. Oliver C, Hyemyung S, Shaida A, Cristina G-L, John G, Maria S, et al. Pharmacological rescue of mitochondrial deficits in iPSC-derived neural cells from patients with familial Parkinson's disease. Sci Transl Med. 2012;4(141):141ra90.

142. Jiang H, Ren Y, Yuen EY, Zhong P, Ghaedi M, Hu Z, et al. Parkin controls dopamine utilization in human midbrain dopaminergic neurons derived from induced pluripotent stem cells. Nat Commun. 2012;3(1):668.

143. Kouroupi G, Taoufik E, Vlachos IS, Tsioras K, Antoniou N, Papastefanaki F, et al. Defective synaptic connectivity and axonal neuropathology in a human iPSC-based model of familial Parkinson's disease. Proc Natl Acad Sci USA. 2017;114(18):E3679–88.
144. Chung SY, Kishinevsky S, Mazzulli JR, Graziotto J, Mrejeru A, Mosharov EV, et al. Parkin and PINK1 patient iPSC-derived midbrain dopamine neurons exhibit mitochondrial dysfunction and α-synuclein accumulation. Stem Cell Rep. 2016;7(4):664–77.
145. Schöndorf DC, Aureli M, McAllister FE, Hindley CJ, Mayer F, Schmid B, et al. iPSC-derived neurons from GBA1-associated Parkinson's disease patients show autophagic defects and impaired calcium homeostasis. Nat Commun. 2014;5(1):4028.
146. Donnelly CJJ, Zhang P-WW, Pham JTT, Heusler AR, Mistry NAA, Vidensky S, et al. RNA toxicity from the ALS/FTD C9ORF72 expansion is mitigated by antisense intervention. Neuron. 2013;80(2):415–28.
147. Di Lullo E, Kriegstein AR. The use of brain organoids to investigate neural development and disease. Nat Rev Neurosci. 2017;18(10):573–84.
148. Raja WK, Mungenast AE, Lin Y-T, Ko T, Abdurrob F, Seo J, et al. Self-organizing 3D human neural tissue derived from induced pluripotent stem cells recapitulate Alzheimer's disease phenotypes. PLoS One. 2016;11(9):e0161969.
149. Tang Y, Liu M-L, Zang T, Zhang C-L. Direct reprogramming rather than iPSC-based reprogramming maintains aging hallmarks in human motor neurons. Front Mol Neurosci. 2017;10:359.

Index

© Springer Nature Switzerland AG 2023
B. Egger (ed.), *Neurogenetics*, Learning Materials in Biosciences,
https://doi.org/10.1007/978-3-031-07793-7

Printed in the United States
by Baker & Taylor Publisher Services